W1FB'S Help for New Hams

By Doug DeMaw, W1FB

Cartoons by Jim Massara, N2EST

Published by: **The American Radio Relay League**
225 Main Street, Newington, CT 06111

Printed in USA

ISBN: 0-87259-443-2

Second Edition

Contents

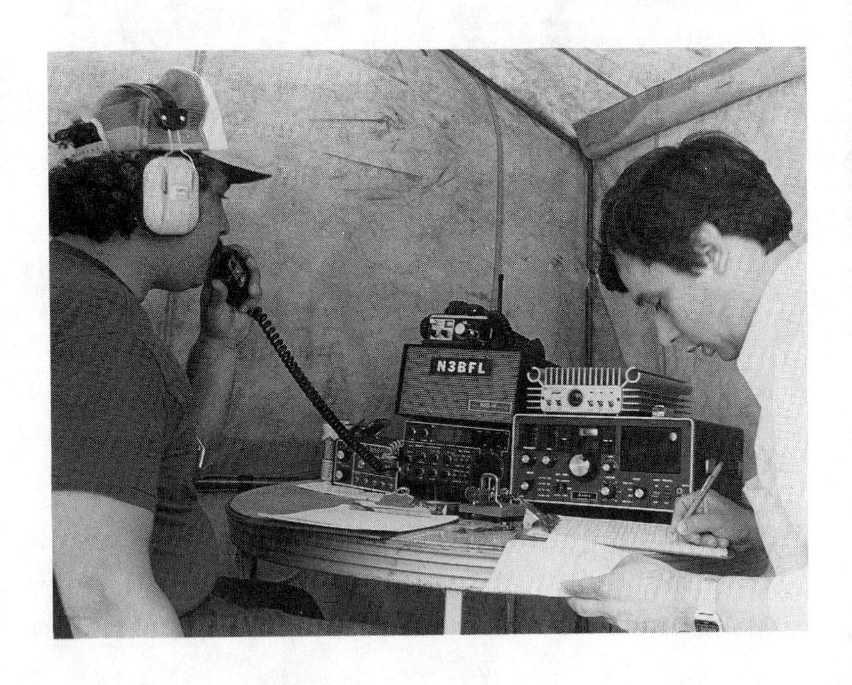

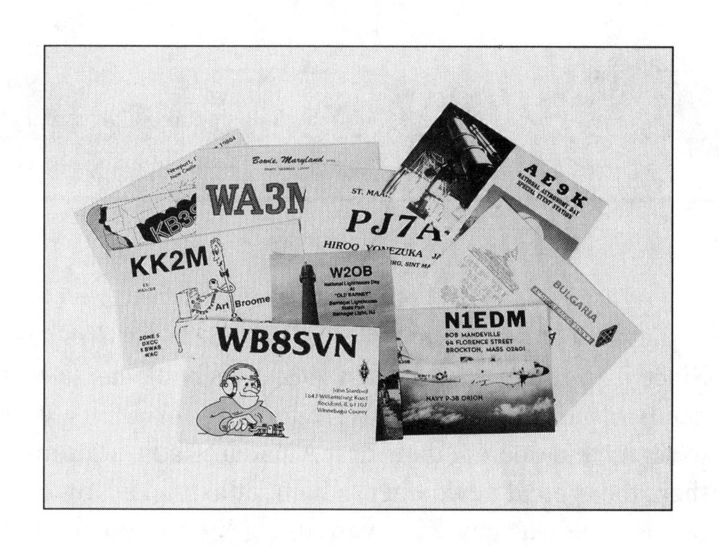
Bowie, Maryland
WA3M
PJ7A
HIROO YONEZUKA
AE9K
KK2M
Art Broome
W2OB
BULGARIA
N1EDM
BOB MANDEVILLE
94 FLORENCE STREET
BROCKTON, MASS 02401
WB8SVN
NAVY P-3B ORION

Foreword to Second Edition

We are proud to present the extensively revised and updated Second Edition of *W1FB's Help for New Hams*. Since its initial publication in 1989, this book has served the needs of newcomers who were looking for advice on how to set up and operate their first Amateur Radio station. Back then, the typical newcomer to ham radio held a Novice class license and was geared toward using Morse code on the HF bands. In contrast, today the overwhelming majority of newcomers are Technicians.

Many Technician class licensees start out on the VHF and UHF bands, on repeaters or packet radio. We've added an overview of the VHF/UHF modes that most newcomers want to learn about, while keeping the rest of the book largely intact.

In addition, the book is now easier to read and use. It has been typeset, photographs have been added and the drawings have been redone.

W1FB's Help for New Hams is but one of an entire library of publications the ARRL publishes to help you get more out of ham radio. For a complete catalog, or for assistance of any kind in getting started in your new hobby, contact the ARRL, 225 Main St, Newington CT 06111-1494, tel 203-666-1541.

There's a handy form in the back to provide us with your comments and suggestions to help us improve the book even further for the next edition. Also toward the back is a form to use to join the 170,000 Amateur Radio operators and others interested in ham radio who are proud to call themselves members of The American Radio Relay League. We would be pleased to welcome you as a League member.

David Sumner, K1ZZ
Executive Vice President

Newington, Connecticut
March 1994

The Amateur's Code

The Radio Amateur is:

CONSIDERATE . . . never knowingly operates in such a way as to lessen the pleasure of others.

LOYAL . . . offers loyalty, encouragement and support to other amateurs, local clubs, and the American Radio Relay League, through which Amateur Radio in the United States is represented nationally and internationally.

PROGRESSIVE . . . with knowledge abreast of science, a well-built and efficient station, and operation above reproach.

FRIENDLY . . . slow and patient operating when requested; friendly advice and counsel to the beginner; kindly assistance, cooperation and consideration for the interests of others. These are the hallmarks of the amateur spirit.

BALANCED . . . radio is an avocation, never interfering with duties owed to family, job, school or community.

PATRIOTIC . . . station and skill always ready for service to country and community.

—The original Amateur's Code was written by
Paul M. Segal, W9EEA, in 1928.—

Chapter 1

Now That You Have Your New License

I *want to get on the air!* Where do I begin? Each of us has experienced the joy and frustration of going on the air for the first time. Take heart; you aren't the first to have the jitters when you uttered your first words into the mike or hit your Morse code key "for real!" But before you deal with these emotional reactions, you should have your new ham station in order and ready to be "fired up." First we'll take a look at VHF radios and antennas, and then we'll cover HF gear.

VHF Equipment

Many newcomers start with a hand-held radio. It's battery operated, so it can be carried anyplace and despite its low power output, it will get you into nearby repeaters or the local packet radio network. It's perfect for the public-service events many hams help with; you can carry your hand-held while assisting with a parade, walkathon or a real emergency. Many hand-held radios can receive outside the ham bands, so you can listen to other types of communication in addition to the ham frequencies.

That brings up another point: Is it best to buy a strictly 2-meter radio or a dual-bander (one that works on 2 meters plus the 222- or 440-MHz band)? The answer may depend on what class of license you expect to have for the foreseeable future. If you're a Technician or you'll be upgrading from Novice to "Tech Plus" (by passing the Technician written exam), you may want a radio that will operate on 2 meters and one of the other bands. Dual-band radios offer greater flexibility because you can always switch to the other band if the local repeaters are busy when you want to talk or ask for assistance. They're also more complicated to use and more expensive than single-banders.

Most new hand-held radios have an array of features: several scanning modes, ability to receive outside the ham bands, up to 100 memory channels and more. Somehow, all these features are packed into a radio that can weigh less than a pound. Accessories include high-capacity battery packs, mobile brackets and telescoping antennas.

Mobile rigs are usually installed permanently (or semi-permanently) in a vehicle. Operating off the vehicle battery, most modern mobile sets have a maximum power output of 40 to 60 W, usually more than you'll need to reach nearby repeaters. (Older mobile rigs on the used market may put out less than 20 W.) Many hams use their VHF radios to talk point-to-point (called "operating simplex") and through repeaters. In fact, it's poor operating practice to tie up a repeater unnecessarily if the station you're talking to is close enough to carry on communications without the repeater. The added output power of a mobile radio allows you to communicate over greater distances than you can with a hand-held rig.

If you want the convenience of a hand-held transceiver with the capability of using higher power in your vehicle or home station, you can buy an amplifier for your hand-held radio. If you hold a Novice license, keep these power limits in mind when you use your mobile or hand-held rig with an amplifier: You can use 25 W maximum on the 222-MHz band and 5 W maximum on the 1270-MHz band.

Where are the Repeaters?

Most ham radio repeaters are owned and operated by radio clubs. Some are operated by individual hams. The *ARRL Repeater Directory* lists repeaters throughout the US, Canada and much of the rest of the Americas, by geographic area and frequency. If you plan to travel with a mobile or hand-held radio, you'll want a copy of this invaluable directory.

Multimode VHF and UHF Transceivers

In addition to hand-held and mobile radios, you can buy VHF gear designed for use at home stations. Many hams use these multimode radios for operating through the ham radio satellites or point-to-point on CW or SSB.

You'll enjoy the challenge and thrills of CW and SSB DXing on VHF and UHF. A multimode (CW/FM/SSB) transceiver is more costly than an FM-only version, but it expands your operating horizons considerably. You'll want to erect a rotatable gain antenna with horizontal polarization when you get involved in VHF or UHF CW and SSB. Most amateurs use horizontally polarized Yagi or collinear beam antennas (arrays) for these modes. Outboard RF power amplifiers are used to provide a signal boost for long-distance communications. A linear amplifier is needed for amplification of SSB energy to prevent signal distortion. Linear amplifiers can be purchased or built from designs in the *ARRL Handbook.*

Antennas for the VHF/UHF Bands

How about antennas for VHF/UHF FM? This is a common question with no simple answer. The greater the antenna gain, the farther away you'll be heard. This isn't necessarily a good feature for an FM antenna that will be used with a high-power transceiver, however, because if your signal is too potent, you can activate more than one repeater that operates on a given frequency. During VHF band openings (tropo or temperature inversions), you might inadvertently turn on as many as five or six repeaters at once. This causes interference to those who try to use the distant repeaters you have no interest in.

The same problem could occur if you operate with more power than you need for your local repeater system. Higher power and bigger antennas are okay for FM simplex operation (point-to-point communications; using the same frequency for transmitting and receiving) when you want to reach a considerable distance.

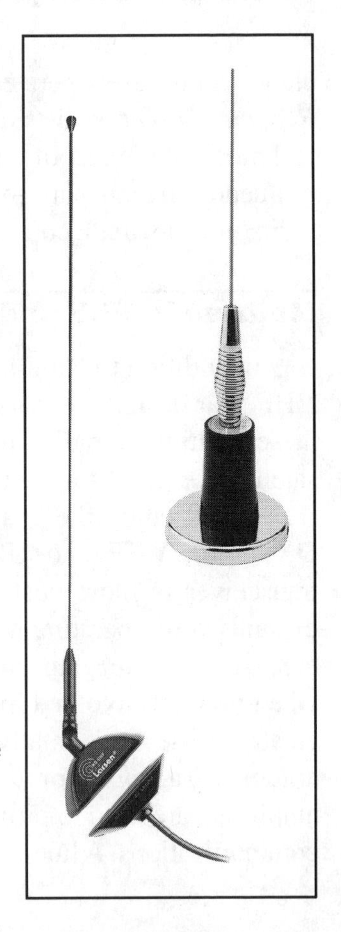

Fig 1-1—VHF/UHF antennas for mobile use can be mounted in any number of ways: through a window, on top of the vehicle, on a bumper or on a trunk.

Every amateur FM band has popular simplex frequencies.

Begin with a simple ⅝-wavelength vertical antenna for home-station and mobile use. These antennas don't cost a lot of money and they provide gain over a ¼-wave ground-plane vertical. Details for constructing a ⅝-wave vertical are provided in the *ARRL Antenna Book* and *W1FB's Antenna Notebook*. You may prefer this low-cost route toward an effective base-station antenna. Alternatively, you can purchase a commercial antenna by one of several well-known, reputable manufacturers. Look at the advertisements in *QST* and visit your local ham retail store for ideas.

For mobile operation, the ⅝-wavelength antenna is preferred for mobile operation, although good results can be achieved with a ¼-wave whip on a vehicle. You can easily make an inexpensive 2-meter ¼-wave mobile antenna from a 19-inch piece of brazing rod or coathanger wire. A similar antenna for 222 MHz is only 12.6 inches long! I recommend the ⅝-wave mobile antenna for best all-around performance. Its angle of radiation is a bit lower than a ¼-wave and it generally boosts your signal by almost 3 dB. That's like doubling your power output (and reception capability) without any additional electronic components! (One caveat: The lower angle of radiation from a ⅝-wavelength whip may have an adverse effect on your signal if you mainly operate in an area where the antennas are far above the average altitude of your station. If you live in a canyon-like area with repeaters on mountaintops high above, you might prefer more RF energy to be radiated at a steeper vertical angle.)

Some amateurs use VHF beam antennas for fixed-station operation. The horizontal beam is the antenna of choice for CW and SSB operation, but is not used for FM repeater operation (the standard multielement beam antenna, such as those used by hams or mounted on roofs of homes for television reception, are known as Yagis, after their codeveloper, a Japanese professor who first described this type of antenna in the 1920s).

Repeaters have vertically polarized antennas; this requires that your FM antenna be vertically polarized. If it isn't, you'll experience substantial signal loss (3 dB or more) because of cross-polarization.

Many commercial Yagi antennas are available for VHF and UHF use, but I encourage you to construct your own antenna. If you use a beam antenna at your home station, use a rotator with it to point the antenna and concentrate its signal energy in the direction of the station you want to work.

HF Equipment

Your selection of items for your first HF station may depend on the money you can invest initially. This shouldn't be influenced by a desire to match the elegance you may have seen at another person's station. It's possible to communicate over great distances with used and simple home-brewed ham gear. It may not be as modern and pretty as that $3000 transceiver you saw in the magazine ads, but it will get you on the air. A 10-year-old Ford can get us from point A to point B as effectively as a new Ferrari. Good operating habits and achievements are more important than expensive radio equipment with all the "bells and whistles."

What About Home-brew Equipment?

If you have moderate technical aptitude, you may want to build some of your first station equipment. *QST*, the *ARRL Handbook*, *Solid State Design for the Radio Amateur*, *W1FB's QRP Notebook*, *W1FB's Design Notebook* and *QRP Classics* (all available through the ARRL) provide transmitter and receiver circuits that can be duplicated by beginners. If you're an electronics engineer or technician, try designing your own circuits for that first ham station. This direct, low-cost approach can get you started with minimum fuss and cash outlay. Besides, this is the only licensed radio service where you're

Fig 1-2—Building your own station accessories is a sure way to enhance your enjoyment of ham radio. This code-practice oscillator is a simple and useful project for those learning Morse code.

encouraged by your licensing authority (the FCC) to tinker and experiment with your own ideas and designs. No review board has to certify your projects for federal approval. As long as they meet a few basic standards, you've got virtual *carte blanche* to twiddle with almost any circuit you can dream up! Whether you buy or build, a relatively modest first station will enable you to evaluate various new and used amateur equipment before you make that bigger expenditure.

Transmitter Power

Maximum legal power (200 W for Novice and Technician/ HF licensees on four HF bands) is by no means essential for communicating with local or distant stations. On the other hand, tackling the crowded Novice and Technician CW subbands with less than 25 W of transmitter output power can make it a challenge to contact distant stations. QRP (low power) transmitters—usually rated at less than 10 W output—are suitable for Novice CW use when band conditions are good, but you won't break through the atmospheric noise and other-station interference every time with just a few watts of radio-frequency power. Because they're lightweight, inexpensive and easy to build, QRP rigs are great for vacationing and camping, but it might take a long time to acquire a Worked All States (WAS) award with a 5-W radio.

The situation is more encouraging when you tackle the Novice 10-meter phone subband or when working 222-MHz FM through a local repeater. Owing to the nature of radio-wave propagation, when the 10-meter band is enhanced by sunspot activity, and because simple antennas perform effectively at 28 MHz, it's possible to earn a spot in the ARRL's DX Century Club (100 foreign countries confirmed) with as little as 5 W! This feat may require a 2- or 3-element Yagi or cubical-quad beam antenna, but it can be done with a vertical ground-plane antenna erected high and clear of nearby conductive objects. In the case of 222-MHz repeaters, if you have a relatively clear path to the repeater you're trying to hit, with no tall buildings or natural features between you and the repeater antenna, you shouldn't need more than 5 W of power.

Some newly licensed amateurs begin with QRP transmitters. Then they get discouraged because they receive few answers to their CQs, they lose communication with stations they're working (because of fading and weak signals) or they're squashed by the signals of louder stations on nearby frequencies. This series of unhappy events has caused some

Novice and Technician US Ham Bands

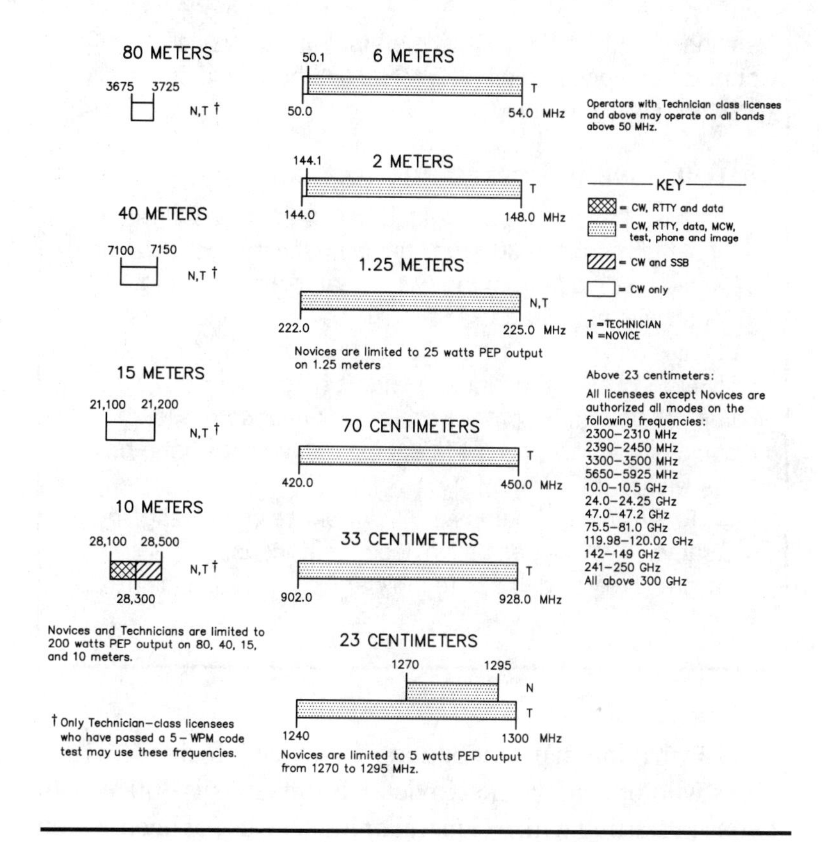

beginners to give up on Amateur Radio altogether. Don't let this happen to you!

In general, a good all-around transmitter output power for a beginner is 100 W. This is typical of most commercially made HF transceivers. This power level will give your signal plenty of "authority" if you have a suitable antenna system.

If You Have a Novice License . . .

...you can make worldwide contacts on three HF bands and you can use Morse code, voice and other modes on another popular HF band, 10 meters. You can use all authorized modes on the 222-MHz VHF band and on the 1270-MHz UHF band.

If You Have a Technician License . . .

...you'll be able to use all the ham bands (and there are a lot of them!) above 30 MHz, including the most popular of all, 2 meters. Aside from the FM repeaters hams use to keep in touch with ham friends and make new ones, the 2-meter band also has most of the packet radio (computer-to-computer) activity. Packet is a means of sending messages and computer data (files, programs, images, etc) across long distances with low-power FM equipment, such as hand-held transceivers.

Technicians who pass a 5-wpm Morse code test gain the same HF privileges as Novices. This license is sometimes referred to as "Technician/HF" or "Technician Plus."

The maximum power output for Novices, Technicians and others who operate in the Novice HF subbands is 200 W. But if you are starting with a General or higher class of license, you may use an amplifier that produces up to 1500 W of output power. You may find an amplifier tempting because it enables you to produce a "smashing" signal that can break the interference/noise barrier. As you learned when studying for your license exam, however, the FCC clearly specifies that amateurs shall use no more power than is necessary to maintain communication. Practically and ethically, you don't need to use high power to talk across town. A maximum-power linear amplifier can increase your 100-W signal by roughly 10 dB. This means that if your 100-W signal is 10 dB over S9, as observed on a signal-strength meter by another operator, it will rise to 20 dB over S9 when you activate your amplifier. Signal levels increase 3 dB each time you double the transmitter output power. The human ear can't detect a 10-dB signal change if the receiver has a good automatic gain control (AGC) circuit, so the practical increase in signal strength may be of no value. On the other hand, if your 100-W signal is at the noise threshold in the other person's receiver (making copy difficult), adding an amplifier can provide Q5 copy (perfect readability). Using an amplifier is a judgment call. You may want to avoid acquiring this expensive item until you've gained significant on-the-air experience. Your power bill will be lower without an amplifier, and you may find that you won't cause interference with your neighbors' TVs, stereos, telephones and electronic gadgets if you stick to 100 W or less.

Used Commercial Ham Gear

You can make your best investment by purchasing a good, used HF transceiver, but you must be wise when choosing a source for this equipment. It can be risky and disappointing to buy an untried rig from a stranger at a hamfest or from a magazine ad. It may have one or more serious defects and you may be unable to find or track down the seller to get your money back. *Caveat emptor* is a good rule of thumb when you purchase used equipment from a person or store you're unfamiliar with. Satisfaction usually results when you buy used gear from a store or person you know. The seller might allow you to try the equipment at his station prior to making the sale or he may be expansive enough to allow you to take the rig home for a short trial before you hand him the cash. Never be bashful about asking to try the equipment before you buy it.

There are many reputable used-equipment dealers from whom you may purchase an older rig. The advertisers in *QST* are required to stand behind their products. If they don't, they can't advertise in the journal. This doesn't apply, however, to the Ham Ads classified section of *QST*. Used equipment from reputable dealers is in operating condition when it's offered. This may not be true of units purchased from private individuals. Several large Amateur Radio dealers stock many used transceivers, receivers and accessories. You may want to send for their flyers before making a choice.

Equipment Features

Specific operating features should be considered before you buy a new or used piece of equipment. Modern ham gear is adorned with features that we can call "bells and whistles." Many options have little or no value to a beginner and the money spent for them can be wasted. For example, you may or may not have a need for computer interface to your transceiver. Similarly, a bank of memory channels may have no application in your shack. In a like manner, dual VFOs aren't essential to most Novice or Technician operation. Your first rig should have the fewest controls possible in the interest of operating convenience. This will help prevent you from becoming confused until you gain confidence and experience. Too many knobs and switches can cause a beginner to wonder, "What if I turn the wrong knob? Will it ruin my rig?" A transceiver with only the basic features enables you to change bands and tune up quickly, and the chance for damage is minimized greatly.

In Chapter 2, we'll discuss studying the equipment owner's/operating manual and practicing transmitter tune-up into a dummy antenna (a 50-Ω resistor that replaces the antenna, but doesn't radiate a significant signal). This is a recommended practice, especially for beginners or people who are becoming familiar with a new rig. Experienced amateurs often say, wisely: "When in doubt, read the instructions."

Let's get back to the bells and whistles for the moment. Older, used ham equipment was built with fewer controls and special circuit features. Some of this gear is easy to become familiar with. This benefit is complemented by reduced cost. It's a good idea to research the performance traits of the gear before making a purchase. Specific items of amateur equipment carry with them a history of performance problems, such as drifting VFOs, final-amplifier tubes that burn out and receivers that overload easily in the presence of strong signals.

Maybe this reads like a shameless plug—and I suppose it is—but the ARRL is rightfully proud of its reputation for performing top-notch Product Reviews. Try to read the Product Reviews of the used gear of your choice, as presented in back issues of *QST* or in the *Radio Buyer's Sourcebook*, a handy collection of *QST* Product Reviews *(Vols 1 and 2* cover radio gear through 1992*)*. You might be able to find back issues of *QST* and other ham magazines in some libraries or you may have a friend who has a *QST* collection. The ARRL's reviewers report the high points and performance shortfalls in their coverage of new rigs. Accurate, thorough, objective Laboratory test results are included, as are subjective operating impressions and observations. There are even comparative reviews that focus on a group of several similar products. Reviews in other magazines may be based solely on the impressions of a nontechnical amateur who tried the equipment for a few days at home. The equipment reviewed in *QST* is always purchased "off the shelf" from regular retail dealers.

Digital or Analog Readout?

There are advantages to having digital frequency readouts on amateur HF transceivers, assuming they can be calibrated against WWV to ensure readout accuracy. Older radios have analog dials that can be off by as much as 1 kHz, even after the tuning dial has been adjusted in accordance with the beat note from an internal calibrator. Also, analog dials provide 1-kHz divisions on the skirt of the main-tuning knob. This coarse resolution is "iffy" for operating near the band edge. A digital readout, on the other hand, offers 100-Hz or better resolution of the operating frequency. This accuracy is helpful when getting on frequency for a schedule with a friend or when calling a net that adheres to a firm operating frequency. It also enables you to operate closer to the edges of a given amateur band.

It's not my intention to vilify older equipment with analog readouts. Many older radios are fine for use as your first rig. The important consideration is that you keep track of the dial calibration and don't stray too close to the edges of your operating frequency band. Check the accuracy of the built-in crystal calibrator against WWV at least once weekly. If your older transceiver lacks a calibrator, use an external one as a secondary frequency standard.

The "KISS" Concept

The "Keep It Simple, Sam" principle is prudent when you select your first rig. We've mentioned the folly of using status as motivation when purchasing amateur equipment. There are older transceivers that are excellent as starter packages. You can upgrade later to a fancier unit, if that's your choice. Your first rig, if purchased used, won't deteriorate in value if you keep it in good operating condition and free of cosmetic blemishes. You should be able to recover all or most of your initial expenditure when changing rigs. Ham radio is similar to hobbies like photography or boating: There's always the temptation to acquire a bigger and fancier piece of hobby equipment, even though your modest unit completely satisfies your fundamental needs. I've known hams who have spent more than $10,000 for radio equipment. Their on-the-air prowess was no greater than other hams who own lower-power, minimum-frills used equipment. It's also ironic that some new equipment has more circuit failures than older gear!

Be on the lookout for well-known older-model rigs in good condition at ham radio flea markets, in stores that sell used gear or in used-equipment ads. In general, avoid buying a transceiver that was built as a kit by another amateur. The builder may have done a sloppy job of wiring the unit, which can lead to all kinds of faults and circuit failures. If you can locate a kit transceiver that was built by an experienced electronics technician or engineer, chances are that the job was done right. Ask to try the rig before making a purchase. It's also wise to take a peek inside to assure yourself that the builder did a decent wiring and soldering job. I've seen kit rigs that looked as "sanitary" inside as factory-wired units.

Station Accessories

Microphones

Most commercial transceivers come with a stock hand-held microphone. Not all of these mikes produce SSB or FM signals that are easy to understand: Some are bassy and others lack sufficient "highs" to provide intelligible transmissions. The effectiveness of your mike largely depends on your voice range. If you have a bassy voice, you'll want a mike that passes high-frequency audio without much attenuation. Conversely, if you have a high-pitched voice (females and youngsters in particular), your mike should have good low-frequency response. Some transceivers have a bass-treble boost control that you can adjust to suit your voice characteristics. Mikes with built-in audio amplifiers have bass- and treble-boost controls, too. Hand-held mikes can produce crunching or grinding sounds on your signal when the push-to-talk (PTT) bar is pressed or held in a closed position. This isn't a pleasant sound for the person at the other end to listen to while you talk! It's better to use a quality desk mike for fixed-station operation. This type of mike is better for voice-operated relay (VOX) operation and the crunching sounds will be avoided during PTT use because you can lock the mike in position.

Amplified mikes aren't necessary, though. Your rig should have sufficient internal microphone audio gain to permit good-quality voice communication without an amplified mike. Microphones with built-in amplifiers can overdrive the speech-amplifier stages in a transceiver, resulting in a distorted SSB signal. Worse still, they're susceptible to malfunction from stray RF energy in your ham shack. The malady appears on your signal as squeals and hum that can make your signal unintelligible. A beginner should avoid using an amplified mike.

Determine the characteristic impedance (in ohms) of the microphone-input circuit of your rig before choosing a mike. Some older transmitters require a high-impedance mike (50,000 Ω) and others are designed for a 600-Ω mike. The wrong mike can cause low transmitter output power or it may accentuate the highs in your voice because of the mismatch. Some modern mikes have a HI-LO switch on them. This allows you to use them with rigs that have high- or low-impedance mike input lines. Older 50,000-Ω mikes may be used with most rigs that have a 600-Ω input if you use an audio stepdown transformer between the mike and the rig to match the impedances. Some amateurs report good results by inserting a 100,000-Ω carbon-composition resistor (¼ watt) in series with the audio lead from the mike. This helps to preserve the frequency response of the mike. An ideal communications mike emphasizes the most useful speech frequencies for clarity. This is the band of audio frequencies between 300 and 2500 Hz. Frequencies above and below this range aren't essential for clear voice communications. I use a modern desk mike that contains a special cartridge that emphasizes the upper voice-frequency range (roughly 2220 Hz). I prefer this microphone because my voice is bassy and difficult to understand when I use the stock mikes that come with transceivers. Experiment with various mikes to learn which is best for your voice.

CW Keys and Keyboards

You have many choices available when you purchase your telegraph key. If you choose a *straight key*, such as the popular war surplus J-38 unit, your maximum sending speed will be restricted to 10 to 20 wpm. This depends on your physical dexterity and natural coordination. The straight key has a fully manual up-down (pump handle) actuation. I've heard letter-perfect CW come from straight keys used by skilled operators. This type of key is perhaps the most trouble-free of the types available to you.

Next comes the "bug" (its physical appearance is somewhat like a huge insect in the eyes of some beholders). This is a semiautomatic CW key. The dashes are made manually, while the dots are automatic up to a certain number. A spring-loaded, weighted key arm vibrates to make a string of dots when the key paddle is pressed on the left side. The bug paddle operates in a left-right motion, rather than up and down. This form of speed key permits a dexterous operator to send at high speeds. Care must be taken to avoid developing what is called a "banana-boat" or "Lake Erie swing" when you send CW with a bug. This kind of CW can be difficult to copy, especially under adverse band conditions: **CQ** comes out as **NN GT**, for example. Also, many bug users don't maintain the correct dot-dash speed ratio. For example, the dots may be sent at, say, 25 wpm while the dashes are only 12 wpm. Adjustment of the bug key can correct this easily.

Electronic keyers are the most popular types of keys available today. The CW actuating device is similar in appearance to a bug key. It's known as a *paddle*. It also has a left-right motion for sending dots and dashes. These paddles may be wired for left- or right-handed operation by switching the leads to the contactors. A paddle is used with an electronic circuit that forms dots and dashes of perfect duration, and correct spacing between the letters. The dots and dashes are self-completing, once you initiate them with your paddle. Electronic keyers don't take care of the spacing between words, though. Proper word spacing is important to avoid causing your sentences to run together like a continuous string of code characters (difficult to copy).

Some electronic keyers have a feature called *iambic keying*. This enables you to use what is known as "squeeze keying," which permits the formation of many of the letters of the alphabet by closing the dot and dash contacts at one time. Such letters as C, K and Q are examples of those that can be formed in this manner. You'll need considerable practice to master the iambic-keying technique. Once mastered, this method allows you to send Morse at faster rates (less wasted motion) than with a non-iambic keyer.

You may prefer to use a keyboard keyer for CW communication. These have a typewriter-style keyboard and a memory that lets you store many code characters as you type in the message being sent. My keyboard stores up to 256 code characters before the buffer (memory) is filled.

The advantage of a buffered keyboard is that you may type well ahead of the message going out over the air. This allows you to make corrections before the incorrectly spelled or chosen words are heard by the other operator. You can walk away from your keyboard (if the buffer is filled) to get a cup of coffee or answer the phone. Your transmission will still continue to be sent.

Fig 1-4—Many hams use a personal computer to generate Morse code. You'll also use your computer for the other popular digital modes, such as AMTOR and packet.

You need not be a touch typist to use a keyboard. I hunt and peck with four fingers, and I use my thumbs on the space bar. I have no trouble sending up to 60 wpm without errors. In fact, I need to be careful to not overflow the buffer when I become enthusiastic about a ragchew conversation!

Some keyboards have additional (smaller) memories that can be used to store short messages for recall as needed. Such messages as **CQ CQ CQ DE W1FB/8 K** or **QRL?** (is this frequency busy?) can be programmed into these short memories for routine operating needs. Those who enjoy contesting find these short memories useful.

If you have a personal computer, you can get software that will let you use your computer as a Morse code keyboard. I've talked to many CW operators who were using computers as keyboards. Software is available for *decoding* Morse code and presenting it line by line on a video monitor. The irony of this type of system is that a person who may not know a letter of Morse code can go on the air and operate CW at 50 wpm! I don't recommend this for beginners. When copying the code electronically, you aren't assured of perfect copy. Signal fading, noise and interference can wipe out entire sentences or cause the screen to display meaningless letters. Poorly sent CW (incorrect letter spacing and defective dot-dash ratios) may appear on your screen as gibberish.

Keyboards are especially helpful to amateurs with disabilities or those with arthritis or similar problems. It may otherwise be impossible to send quality CW with a straight key, bug or paddle.

I switched to a keyboard after age 55 because my brain-to-arm coordination declined to a degree that prevented me from sending error-free CW consistently at speeds above 30 wpm. My ability to copy speeds up to 50 wpm in my head hasn't declined, however.

Other Accessories

Another station accessory to consider is an antenna tuner (transmitter to feed-line matcher). These are also called *ATUs* (antenna tuning units), *antenna couplers* and *Transmatches*. Some transceivers have automatic, built-in tuners, and these often do the job. Tuners consist of coils and variable capacitors that can be adjusted to provide a resistive 50-Ω load for your transmitter and receiver. If your antenna system presents a high standing-wave ratio (SWR) to the transmitter, the built-in SWR-protection circuit in your rig will cause the transmitter output power to decrease. Adjustment of the antenna tuner will provide the desired 1:1 SWR the transmitter needs for normal operation.

Antenna tuners are necessary when you use multiband end-fed or dipole antennas that have tuned feed lines. They shouldn't be required when you use properly matched coaxial-fed dipoles and beam antennas. The exception is when a dipole or beam antenna is tuned for low SWR in a given part of an amateur band, but exhibits a high SWR somewhere else in the band. This may happen when you try to use an 80-meter dipole for the 75-meter phone band or vice versa. The antenna tuner can't correct the mismatch at the antenna feedpoint, but it will ensure a low SWR for the transmitter at that end of the feed line.

Your antenna tuner must be capable of accommodating the output power of your transmitter without arcing and overheating of the components. Check the ratings of the unit you buy and make certain it's suited to your transmitter power. Specifically, high-power antenna tuners should have wide spacing between the plates of the capacitors and the tapped coil or roller coil needs to have wire of large diameter (gauge) to minimize losses from coil heating. The switches need to be made with large contacts and ceramic insulation to handle high power.

If you use an antenna tuner without a built-in SWR meter, you'll need a separate SWR meter to use with it. Placed between the transmitter and the antenna tuner, the meter lets you know when the antenna tuner is tuned for minimum SWR as you observe the reflected power on the meter. All tuning is done for minimum reflected power, which coincides with a 50-Ω impedance match. You can build a homemade SWR meter or bridge for moderate cost. The *ARRL Handbook* and the *ARRL Antenna Book* provide circuits for simple SWR indicators. You can save money, while having fun, if you build your own. Avoid using SWR meters designed for Citizens' Band equipment. Many of these instruments are poorly designed and don't work well below 27 MHz.

Antennas for the HF Bands

I won't attempt to be all-inclusive in this discussion about HF-band antennas. To cover this subject from cellar to penthouse it would be necessary to fill several pages of this book. The best advice I can offer while being terse is that you adopt the KISS principle mentioned earlier. It's difficult to beat the low cost and simplicity of a ½-wave dipole antenna. Fed with coaxial cable (except when tuned feeders are used), they provide good performance if they're erected high and in the clear of nearby conductive objects (power lines, phone wires and metal structures). Keep your dipole away from trees when possible and don't allow the antenna wires to lie on the tree limbs if you must route it through trees or brush.

Dipoles are easy to construct, but you can purchase them if you prefer. I encourage you to build simple antennas instead of buying them. You'll save money and gain technical knowledge and skill by making your own antennas.

Some amateurs erect a separate dipole for each band of operation. You may lack sufficient real estate for this. You can solve the problem by putting up a multiband trap dipole that has a single coaxial-cable feed line. These dipoles have a number of traps on each side of the feedpoint. In effect, the traps block the passage of RF energy at certain frequencies, while allowing it to pass the desired frequency. You might think of these radiators as "smart antennas." On some design frequencies, you use only a small part of the dipole, but on other frequencies the entire dipole is the radiator. This principle also applies to multiband beam and vertical antennas.

Another scheme for using a dipole on several ham bands is to connect two or more sets of dipole wires to a common feedpoint. For example, you might construct an antenna that has dipole legs or wires for 80, 40, 15 and 10 meters. The individual dipoles are fanned so that their outer ends are spaced from one another. A single 50-Ω or 75-Ω coaxial feeder is used with this system. Multiband operation can also be accomplished by cutting a dipole to the proper length for 80 meters and feeding it with a tuned, parallel feed line, such as 300-Ω TV ribbon, 450-Ω ladder line or homemade open-wire line. In this manner, it's possible to use the basic 80-meter dipole for the remainder of the amateur HF bands by placing a balun transformer and antenna tuner between the feeder and the transmitter.

A straight horizontal dipole requires two support poles (one at each end of the antenna). You can use a single support pole by changing the dipole to an inverted V. In this example, the center (feed point) of the antenna is attached to the supporting device, but the ends of the dipole are drooped to a few feet above ground. The end insulators can be guyed to a fence, bush, tree or short wooden stake. The best enclosed angle for an inverted V is approximately 90°. This provides omnidirectional radiation and vertical polarization. This style of antenna is also known as a drooping dipole.

Vertical antennas provide excellent results when installed correctly. A ¼-wavelength vertical antenna requires a ground screen or groundplane under it if you're to enjoy optimum performance. A ground-mounted vertical has radial wires fanned out from its base. Some hams use as few as four radials and others use as many as 120 wires, laid out as spokes on a horizontal wheel. The more radial wires you use (up to 120), the better the antenna efficiency. The wires can lie on the ground, be buried a few inches in the soil or they can be fanned out above ground and supported by short poles a few feet above ground. Some commercial verticals don't require any radials.

A full-size, single-band vertical yields the best performance. Short, trap vertical antennas can be purchased if you're willing to accept a performance compromise (narrow SWR bandwidth and lower efficiency). This type of multiband antenna is often the choice among hams who have limited antenna space.

In Chapter 3, we cover this subject in considerable depth. Suggested reading material for this subject is contained in the *ARRL Antenna Book* and *W1FB's Antenna Notebook.*

Using the Exotic Modes

It can be argued that "exotic" is a misnomer when we refer to special modes available to hams. After all, as radio amateurs, we're skilled in our craft and no popular communications mode is beyond our comprehension. Among the exotic modes are amateur television (ATV), which includes slow-scan TV (SSTV) and fast-scan TV (FSTV); radioteletype (RTTY), which includes Baudot, ASCII and AMateur Teleprinting Over Radio (AMTOR); digital computer data communications, such as packet, PacTOR and Clover; microwave communications; and Orbiting Satellites Carrying Amateur Radio (OSCARs). You can use facsimile (fax) and even more esoteric modes. These include earth-moon-earth (EME or moonbounce), meteor scatter for long-distance VHF work, and even spread-spectrum UHF/microwaves.

You may find this a bit mind-boggling at first, but things will be less confusing as you gain experience in Amateur Radio. A thorough explanation of each of these modes is beyond the scope of this book. You can get detailed information by referring to the *ARRL Handbook*, the *ARRL Operating Manual, Your RTTY/AMTOR Companion, Your VHF Companion, Your Packet Companion, The UHF/Microwave Experimenter's Manual, The UHF/Microwave Projects Manual* and the *Satellite Experimenter's Handbook*. I suggest you avoid becoming involved with more than one of these operating techniques at a time. Special equipment is needed and it will require several weeks to feel "at home" when you sit down to operate. Keep in mind the KISS principle we considered earlier.

You'll have opportunities to experiment with equipment for the amateur microwave frequencies when you receive your Technician license. Being an active microwave pioneer places you in a unique group of hams with unusual technical skills. Microwave equipment is fun to build because it's similar to assembling plumbing fixtures. Sheets of metal and tubing provide the foundation for microwave gear, and electronic circuits can be scavenged from microwave ovens, surplus military and industrial equipment, and even automobile radar detectors. You can combine projects in projects, such as a high-speed microwave packet system (for example, see pages 32-4 through 32-11 in the 1994 edition of the *ARRL Handbook*). Maximum pride and satisfaction can be yours if you try your hand with the special modes that make Amateur Radio so fascinating.

Summary

Your first steps in Amateur Radio can be frightening without an "Elmer"—an experienced ham who can get you started. I hope to serve that purpose for you through this book. This and the chapters that follow can help you avoid the fright and indecision new hams often experience. In this chapter, I've tried to cover questions new or aspiring hams are likely to ask. Here are some general guidelines to follow:

1) KISS

2) Start modestly and grow slowly.

3) Avoid buying a new or used rig that has bells and whistles you may never use.

4) Avoid basing your equipment needs on status.

5) Operate only a few modes (FM, SSB and CW) at first. Avoid the exotic operating modes until you gain experience.

6) Build your own simple antennas.

7) Avoid buying unnecessary station accessories.

8) Read *QST* product reviews in the *Radio Buyer's Sourcebook* concerning gear you plan to buy.

9) Study your gear's operating manual before hitting the ON switch.

10) Buy your equipment from reputable amateurs or dealers who will guarantee the performance.

11) Join the ARRL and receive *QST* each month. This will keep you apprised of new equipment features, FCC rules changes and current operating practices. *QST*'s New Ham Companion section is designed for beginners.

13) Obtain a current edition of the *ARRL Operating Manual*, the *FCC Rule Book* and the *ARRL Handbook.*

14) Have fun…this is a spare-time hobby or service, however you want to classify your activity.

Chapter 2

Your New Equipment—Getting Acquainted

That new amateur license looks impressive! You're proud of it, and rightfully so. Make a photocopy of the ticket, frame it and display it on the wall in your ham shack. The original license should be filed away and kept secure. Alternatively, you may elect to frame the original document and file a photocopy for future use. Once this formality has been taken care of, you're ready to face that collection of new and perhaps mystifying array of station gear.

There's no need to be timid about "shaking hands" with your station equipment. Follow the order of things as they're listed in this chapter and you'll grasp the operating fundamentals with ease. Don't allow your station gear to intimidate you;

we all had to start at the beginning at some time in our amateur careers. I have yet to meet a ham who was defeated by the apparent complexity of adjusting an antenna system or recognizing an equipment malfunction. Your collection of radio equipment isn't a specter in waiting.

Your Friend, the Operating Manual

Hams have a profound saying that you should adopt as part of your ham philosophy: When in doubt, read the manual. The operating book for your equipment is the best source of "how-to" information you have at your command. Many hams are tempted to give these important books a cursory examination and set them aside in favor of groping for the proper control settings. Read your instruction book three or four times before you flip the power switches on your gear. Take notes as you read the manual and pay particular attention to the parts that tell how to connect the cables. Check cables for opens or shorts—use a volt-ohmmeter (VOM) for these tests. A faulty cable can cause endless grief for a new ham trying to put his equipment on the air for the first time.

A notebook with index tabs is useful when you want to refer to notes you've taken from the equipment manual. Your tune-up notes and other pertinent data will help you look up a specific procedure without thumbing through the manual or reading paragraphs that don't contain the information you're seeking. I keep notebooks at hand for my transceivers and amplifiers. The notebook tabs are useful for categorizing the data, such as Gain Settings and Filters.

After You Study the Manual

After you compile your notes, it will be time to practice tuning your rig. During this process, you'll memorize the adjustment procedures and this will give you the confidence you need when you go on the air. Practice tuning the equipment with a dummy antenna (a 50-Ω resistive load), rather than putting a signal on the air. You don't want to interfere with stations operating near the frequency you've chosen. A dummy antenna allows your transmitter to connect to the impedance it's designed for (50 Ω), whereas your antenna may not present a 50-Ω load without being matched for that impedance. A severe mismatch, or one that causes an SWR greater than 2:1, can disrupt the operation of your transmitter and cause problems during the tuning process. In a worst-case situation, high SWR can damage the final amplifier devices in some transmitters.

A dummy antenna is a power resistor that dissipates the transmitter's RF output power as heat, and is air cooled or suspended in oil. It's a purely resistive 50-Ω component that isn't frequency sensitive across the range of MHz for which it's built. "Purely resistive" means that there's virtually no stray inductance or capacitance (abbreviated X_L and X_C, respectively). Unwanted capacitance or inductance can cause the dummy antenna to be reactive (X) and this prevents it from presenting a true 50-Ω resistance. The condition becomes worse as the operating frequency is raised, say, from 7 to 21 MHz. A reactive dummy antenna that contains a 50-Ω resistor can, for example, present a 100-Ω or even a 25-Ω load when connected to your transmitter.

A dummy antenna doesn't radiate much RF energy if it's well shielded in a metal enclosure. You can connect an earth ground to the shield can or box to further reduce unwanted radiation while you're testing your transmitter. A dummy antenna is one of the first pieces of accessory gear you should acquire. You'll need to use it often. Amateurs who tune up with their signals on the air may elicit remarks such as, "That guy has his dummy connected to the wrong end of his rig," meaning that the dummy is at the transmitter's controls. Do your initial tuning into a dummy antenna! Final touch-up can be done quickly with the antenna attached to the transmitter.

Most modern VHF/UHF FM transceivers don't require or even provide external controls for tuning up. They simply offer a 50-Ω antenna jack and expect you to plug it into a 50-Ω feed line and antenna system. Hand-held transceivers include a "rubber duckie" style antenna as a nearly perfectly matched device. Mobile transceivers provide a variety of outputs: a single-band jack, and dual-band jack that must be plugged into a dual-band antenna, or a pair of single-band output jacks that each plug into separate antennas or may be used with multiband antennas when attached through a device called a *diplexer.* The diplexer accepts signals from all of the radio's bands and provides separate output connectors for single-band antennas. A *triplexer* may be needed for a three-band rig if you plan to use three separate antennas. Some transceivers have built-in circuits to permit you to plug in multiband antennas without any external boxes or adapters. A duplexer can be used to combine separate outputs from a dual-band transceiver if you want to use a single multiband antenna. Check the manufacturer's literature or the owner's manual for details. As with HF, a dummy antenna is a useful, inexpensive device for testing VHF and UHF transmitters without putting unnecessary signals on the air.

Tuning Up a Tube-Type Transceiver

You won't find a front-panel arrangement that can be called typical when you look at commercial transceivers. Each brand and model may have controls for features not found on other amateur gear. Because many beginning hams start out with older equipment that they buy used, or borrow from an experienced ham, a general front-panel layout for a tube-type transceiver is shown in Fig 2-1. It's patterned somewhat from the Yaesu FT-102 transceiver I use. We'll go through a practice tune-up based on Fig 2-1.

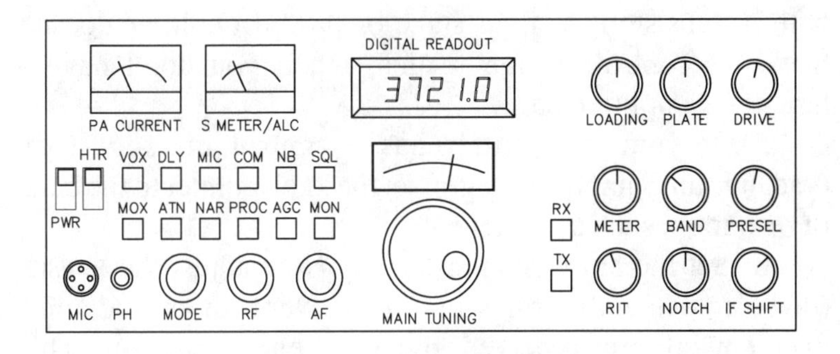

Fig 2-1—A pictorial representation of the front panel of a tube-type HF transceiver. The square controls are push buttons and the circular controls are standard knobs.

First, connect a dummy antenna to the antenna jack of the rig. This requires a short length of 50-Ω coaxial cable. Plug the ac cord into a wall outlet. Make sure the MOX switch is in the OFF (VOX) position. Turn on the POWER switch. An older radio may have two such switches, as in Fig 2-1. The second one is labeled HTR (heater). This extra switch is only found on equipment that uses vacuum tubes. It turns on the tube filaments. A hybrid transceiver that contains semiconductors (transistors and integrated circuit chips) and tubes may have an HTR switch. It's kept in the OFF position when you're only using the receiver portion of your rig. In this practice run, we'll actuate the HTR switch and the POWER switch to permit the transmitter to operate:

1) Set the DRIVE (sometimes called CAR. LEVEL) control to minimum (counterclockwise).

2) Select a test frequency, such as 3700 kHz.

3) Set the meter switch for reading plate current (IP) or power amplifier current (PA).

4) Position the LOADING control to read zero. Equipment with a transistorized final stage or power amplifier doesn't have this control, so ignore this step if you don't have a tube-type amplifier stage.

5) If your transmitter has a control marked PLATE (vacuum-tube amplifier stage), set the knob pointer to the band of operation, such as 7 MHz.

6) Set the MODE switch to CW. Now adjust the PRESEL (preselector) knob for maximum receiver background noise or for maximum received signal, if one is present. This setting corresponds to the proper setting for maximum transmitter output power. Ignore this step if your unit has no PRESEL control.

If some of these steps refer to controls and switch positions that aren't found on the front panel of your rig, check your operating manual to learn which control or switch position corresponds to the function names used here; not all manufacturers use the same names for the controls.

7) Set your MODE switch to the TUNE or CW position. If you use the CW position, you may have to connect a CW key to the key jack to actuate the transmitter (key closed). Next, place the MOX/VOX control in the MOX position. Some rigs have a PTT (push-to-talk) label instead of one that indicates MOX. If so, use the PTT position. This will require that you use your mike switch to actuate the transmitter. Using the MOX mode shouldn't require the use of a mike or key for tune-up.

8) Turn on your transmitter. Observe the IP or PO (relative output power) scale on the panel meter. It should read zero or nearly so. Slowly advance the DRIVE control until the meter reads approximately ⅓ scale. Now adjust the PRESEL control for maximum meter reading. You may now adjust the DRIVE control for the rated IP of your transmitter.

9) Next, you'll have to adjust the PLATE and LOAD controls, in addition to the foregoing steps. This must be done at reduced power to prevent tube damage. The DRIVE should be kept low

enough to restrict the I_P reading to roughly ⅓ the full rated amount. Don't keep the transmitter keyed for more than 30 seconds without allowing a 30-second cooling period. This prevents wear on the tubes. This applies to normal operation and to the tune-up period.

10) With the transmitter activated, observe the I_P meter reading and adjust the PLATE control until you see a sharp dip in meter reading. The needle swings to the left when this occurs. Adjust the PLATE control for the lowest meter indication. This corresponds with resonance of the tuned output circuit at the chosen operating frequency. Next, adjust the LOAD or LOADING control slightly to cause an increase in I_P reading. Readjust the PLATE, tuning for a dip in I_P. Repeat this process until the dip in I_P is fairly shallow or about 80% of the maximum I_P reading before the dip is found. You now have your transmitter properly loaded and you can advance the DRIVE for the recommended I_P, as set forth in the operating manual. Repeat the tuning process each time you change operating frequency.

11) You can use the P_O metering mode for touching up your PRESEL, PLATE and LOADING controls after you've finished the basic tune-up process. The meter will indicate RF power output. Set the knobs for maximum P_O indication, consistent with the recommended I_P for your transmitter. Don't exceed the proper I_P to squeeze out a few more watts. This will shorten the life of your final amplifier and may cause the tubes or transistors to be destroyed.

12) Always tune up with a dummy antenna connected to your transmitter. Once you've optimized the settings of your controls, you can switch to the station antenna and perform a short final tune-up. It's easy to install a switch in your shack that allows you to change quickly from the antenna to a dummy antenna. All you need is a single-pole, double-throw (SPDT) ceramic rotary switch. It can be installed in your antenna tuner if you prefer to have things tidy in your shack. This requires

the addition of a coaxial jack (for the dummy antenna) on your tuner.

13) The antenna may present a different load to your transmitter, compared to the resistive 50-Ω dummy antenna. This generally dictates the need to touch up the PLATE and LOAD settings for a vacuum-tube amplifier stage. The difference you'll observe when connecting the antenna to your transmitter is caused by X (reactance) in the antenna system. It's unlikely that your antenna will present a 50-Ω resistive (flat) termination for your transmitter. There's always some SWR present, however low the ratio, caused by an imperfect match between the antenna and its feed line. As your operating frequency is moved away from the resonant point of your antenna, the X will increase, thereby increasing the SWR.

Tuning Up a Solid-State Transceiver

Most modern solid-state transceivers don't have tuning controls. They are equipped with broadband output networks that automatically band-switch as the VFO is tuned to different frequencies. However, many early solid-state transceivers were not equipped with this capability and have a band switch (BAND) and/or a preselector adjustment (PRESEL). If you have one of these rigs, consult your owner's manual for the proper tuning procedure.

SWR and the Solid-State Transmitter

Modern transistorized amateur transmitters contain built-in SWR protection circuits that sense the reflected power from your antenna. The greater the SWR, the lower the output power from the transmitter. The sensing circuit supplies a reverse bias voltage to the driver stage, which lowers the power output of that stage. This reduces the driving power to the final stage (amplifier) of the transmitter to reduce the amplifier's dc current, protecting the amplifier from burnout or excess heating. This feature makes it necessary that you match your feeder to the antenna or use an antenna tuner to ensure that the feed line presents a 50-Ω load to the transmitter. The tuning network in a vacuum-tube amplifier permits you to match the output amplifier to the antenna system over a limited range of impedances above and below 50 Ω. Many newer solid-state rigs have automatic tuners built in.

Dial-Settings Chart

Draw up a chart that shows the approximate settings for your transmitter dials for each band you operate. You may need two sets of numbers on your chart for 160, 80 and 40 meters (for the high and low ends of each band). A tuning chart is a great convenience when you change bands or operating frequencies within one of the lower-frequency bands. Include dial settings for your antenna tuner, too. Finding the proper knob and switch positions without a chart can be time consuming. This is particularly true when operating in the 75, 80 and 160-meter bands with tube-type transmitters because the PLATE control is usually marked in MHz. These frequency marks can be meaningless with respect to the actual operating frequency. My transceiver, for example, has the PLATE tuning knob set at 20 MHz when I operate 1.9 MHz. Don't let this condition upset you if your rig works the same way. Your chart will be beneficial, especially for tune-up on the lower frequencies in the HF and MF spectrum.

If you become a contest operator or DX chaser, you'll often need to change bands or frequencies. A good calibration chart will be your best friend. Final tune-up of the rig is a simple matter once you orient the controls for the coarse settings on your chart.

You can simplify matters even more by adopting the "o'clock" method on your chart. For example, if a given knob pointer is straight up for proper operation, make your chart notation as "12 o'clock." If the pointer is 90° to the right, call it "3 o'clock."

CW Operation

Many amateurs prefer to operate CW with the transmitter set up for break-in delay (VOX). The VOX circuit has controls you can adjust (internal or external) for VOX GAIN and VOX DELAY. The gain control must be advanced until a "dit" from your key causes the VOX relay to actuate, which activates the transmitter. The VOX DELAY control is then set for the dropout time (delay) you prefer. If you allow too great a delay (relay release), you may miss a few code characters from the other station when you stand by. This is because the receiver doesn't turn on until the VOX relay releases. Too little delay will cause the relay to cycle each time you close your CW key. I like to set the delay in my rig for approximately one second. The delay you decide on will be a matter of personal choice.

Some modern transceivers feature instant break-in (QSK). When using this system your receiver recovers instantly when you release the key. QSK is also called "full break-in." Ragchewers and those who enjoy handling traffic (messages) find QSK especially delightful because of the fast action it permits.

If you don't care to use break-in delay or QSK, you may set the switch for MOX during each transmission. This keeps the receiver silent until you switch from MOX to VOX during receive. The MOX mode locks the changeover relay in the transmit position until you release it by switching back to VOX.

Sidetone Level

Your transmitter probably has a *sidetone* circuit. This is used to monitor your CW sending to help prevent errors. The tone from an audio oscillator inside your rig is heard in your headphones or loudspeaker each time you close the CW key. There should be a potentiometer inside your transmitter to adjust the sidetone volume—check the manual. If you use an electronic keyer, keyboard keyer or computer to send code, there's probably a provision for a sidetone oscillator and speaker to sound from it. If so, you can turn off the sidetone output in the transceiver and use the tone from your keyer, or vice versa.

SSB Operation

Adjusting your transmitter for SSB operation is more critical than for CW. Particular attention needs to be paid to the mike GAIN setting. Too much gain causes audio distortion on your signal and can cause RF distortion in the transmitter. The latter condition may create an excessively wide signal that occupies many kHz of the spectrum and interferes with conversations elsewhere in the band. This is against FCC regulations, and it irritates other operators! Excessive audio and distortion is rampant on the phone bands. In most instances, it can be cured by simply turning down the mike gain. Human nature seems to indicate that the higher the mike-gain setting, the better you'll be heard. The reverse is true: A distorted signal is difficult to copy.

The correct setting for the mike gain of your rig depends on (1) your voice characteristics and (2) the output level of the mike you use. Finding the correct setting requires experimenting. Too little gain will cause your transmitter output power to be low and too much gain will make your audio sound rough. If you know how to use an oscilloscope, you can observe the transmitter's output waveform and adjust the audio level for maximum output power, consistent with minimum signal distortion. Otherwise, your best course of action is to engage in an SSB QSO and have the other operator assist you in setting the mike gain for the best signal quality and minimum signal

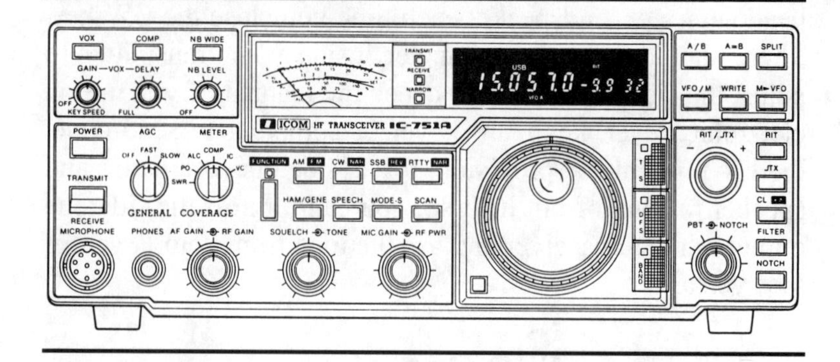

Fig 2-2—The front panel of a modern, all-solid-state transceiver. Operating conveniences include adjustable noise blanker, general-coverage receiver, dual VFOs, memories and scanning.

bandwidth. Most hams are willing to lend a hand in this manner.

Your operating manual should have tips for setting the mike gain. You may find that there's a recommended meter reading on the automatic limiting control (ALC) scale for audio peaks. If so, use this information when making your initial mike gain adjustment. Subsequent on-the-air adjustments can hone the signal quality.

If you have a second receiver available, remove the antenna from it and attach a 12-inch piece of wire to the antenna post. You can use this receiver to monitor your SSB signal as you adjust the mike gain. Wear headphones for this test to help block out your voice as you speak into the mike.

Mike Impedance

It's essential that you use the correct mike with your transmitter. Check the operating book to learn the impedance of your mike input circuit. Most modern rigs call for a low-impedance (600-Ω) mike. Tube rigs and some older solid-state transmitters call for a high-impedance mike (about 50 kΩ). An incorrect mike can result in low transmitter output power and excessive high-frequency audio response.

What about Speech Processing?

Processors have their advantages under certain conditions, if adjusted properly. Unfortunately, this seldom seems to be the case. Many amateurs are tempted to use excessive processing and excessive mike gain. The needle on the IC meter flops vigorously and gives the operator the feeling that he's "really getting out!" A processor is supposed to increase the average output power of the transmitter. Specifically, the softly spoken syllables are boosted to a level that causes greater transmitter output power than without processing. The peak (maximum) output power of loudly spoken syllables remains about the same when using processing. Processing is rated in dB of speech compression. If your transmitter meter has a scale for compression, keep the maximum needle deflection below 6 dB. This setting gives "presence" to your voice signal without causing you to sound like you're speaking from the bottom of a well or barrel. Your mike gain needs to be adjusted (with the processor on) for an undistorted output signal.

One bad aspect of a speech processor is that a mike circuit is more sensitive with the processor than without it. This causes all kinds of extraneous sounds to be heard over the air. A dog barking elsewhere in the house, sounds from the next room and the noise from your radio's blower fans will be heard plainly along with your voice. This doesn't aid communication; it degrades it. Your objective during voice operation should be to sound as natural as you can. Processors prevent this from happening and they usually add nothing worthwhile to your signal. Finding the correct combination of mike gain and processing levels is no easy assignment. It calls for using a 'scope or a trained ear at the other end of your communication circuit.

When you use your speech processor, there's an increase in the average power of your transmitter. This means that it must work harder than when no processing is used. The greater the compression level in dB, the more severe the duty cycle of the transmitter. This effect can shorten tube life if carried to excess and this includes the tubes in a linear amplifier driven by a rig with speech processing. Avoid using your processor for routine communications. Be judicious; if your signal is weak and the copy is marginal, your speech processor can make the difference between poor copy and solid communications.

SSB and Voice-Operated Relay (VOX) Controls

Adjust your vox gain control for close talking into the mike. The proper distance between your mouth and the mike is about 6 inches. This will help prevent unwanted reproduction of your breathing and should eliminate "blast" noise from your lips as you speak. Set the VOX control so the VOX relay engages when you speak into the mike with a normal voice level. Don't shout! Adjust the VOX DELAY for a dropout time of roughly 1 to 2 seconds. Too long a delay will result in lost words at the start of the other operator's transmission. Too short a delay time will cause the VOX relay to cycle excessively. Be sure the VOX GAIN is advanced far enough to prevent "clipping" the first syllables of your sentences as you speak.

Your rig probably has an ANTIVOX control. It's adjusted to prevent the audio output (speaker) of your receiver from actuating the VOX circuit. The symptom of an improperly adjusted antiVOX is a VOX relay that comes on and off (clatters) when you're talking into your mike because the receiver audio is picked up by the mike and transmitted.

Locate your mike as far away from the speaker as possible. This lets you adjust the antiVOX circuit more easily. You shouldn't have any problem if you wear headphones during VOX operation.

If you don't want to use VOX during voice communications, you can use push-to-talk (PTT) operation. Your transmitter will operate only when you press the PTT switch on your mike. I prefer PTT operation because it prevents extraneous sounds in my house from activating my VOX circuit. Sneezes, coughs and clearing your throat won't go over the air if you use PTT. The shortfall of using PTT is that the other operator must wait until you release your mike switch before she can speak. VOX operation is better for roundtable chats.

The Woes of Transmitter Misadjustment

What can happen if you misadjust your transmitter? Some consequences are (1) Low RF output power; (2) excessive RF output power; (3) emissions outside the chosen amateur band and (4) damage to the transmitter. Too much power output from your transceiver can damage your linear amplifier, if you use one.

Too much drive power to the last stage of your transmitter can cause spurious energy to be transmitted. The amplifier can become nonlinear when driven too hard. This generates unwanted harmonics that can cause television interference (TVI) and radio frequency interference (RFI). This spurious energy can disrupt reception in nearby home entertainment systems, TVs, VCRs, telephones and broadcast receivers, and may affect local commercial two-way radio systems. A misadjusted transmitter can cause a wide signal that interferes with the communications of your fellow amateurs within a given ham band. This and the transmission of spurious signals can cause you to be cited by the FCC (which may issue you a pink ticket, as the notice is called).

During CW operation your transmitter may cause the CW note to be "clicky" if you exceed the recommended operating current of the equipment. Key clicks spread transients across a ham band and even outside the band. This can earn you a pink ticket and won't make you a hero in the eyes of those who share the band with you.

I've known amateurs who believed their signals would be heard farther away if they operated their transceivers beyond the rated power limits. They also had a penchant for shouting into their mikes as they talked, hoping to produce a louder signal. Ironically, this can degrade, rather than improve your signal. Follow the instructions in your operating manual and you should produce a clean, easy-to-copy signal. Pay special attention to the tune-up procedure and meter readings.

If your transmitter puts out poor signals, you may be fortunate enough to be monitored by an Official Observer (OO). An OO is a trained ARRL Member who's also a member of the FCC's Amateur Auxiliary corps. Although OOs play a role in assisting the FCC under certain circumstances, such as helping gather evidence for a formal investigation, the OO's primary goal is to help you with off-the-air advice and observations. An OO is not some kind of "ham cop." The OO's job is to monitor his fellow hams and take note of on-the-air difficulties—and of exceptionally good operating practices. If he hears the former, he may mail you an Official Advisory postcard describing the incident and what type of transgression you may have committed. If your operating skill and transmission quality is exceptional, you could receive a Good Guy/Gal Report. The OO program is an excellent opportunity for amateurs to help amateurs, because an Advisory Notice carries no legal weight, but helps you take steps to correct or prevent a problem from recurring before the FCC hears it and issues a formal citation. Astute hams appreciate an OO's friendly advice and praise the service these volunteers perform on behalf of their fellow amateurs.

Transceiver Controls that Affect Reception

Receivers have a wide array of controls to adjust. This intimidates some new amateurs. There's no need to fear the receiver adjustment process. Just as no two brands or models of transceivers have the same controls or control labels, the same is true for receivers. Assuming you'll recognize the controls I mention, even though your unit may have them labeled differently, the controls mentioned here should be similar enough for you to follow this discussion.

RF Gain or Attenuator

The *RF amplifier* is the first stage in your receiver. It amplifies incoming signals from the antenna before they reach the remainder of the receiver circuit. The second stage is called the *mixer*. Your transceiver may have a control marked RF AMPLIFIER, RF GAIN or PREAMPLIFIER. This control switches in or removes the RF amplifier. The incoming signal goes directly to the mixer when the RF amplifier is switched off. This causes a drop in received signal level and you'll see your S meter reading fall at the same time. This generally amounts to a drop in gain of about 10 dB.

The front-end ATTENUATOR control has the same effect. It consists of a resistive network that presents a 50-Ω impedance to the first stage of your receiver. Depending on the receiver design, the attenuator may decrease the signal by 10 to 20 dB. You're probably wondering what the value might be for an attenuator or switchable preamplifier. These devices help prevent receiver overloading when strong signals are present. Imagine that you have a neighbor who's a ham operator. If he decides to operate on 40 meters at the same time you're using the band, his strong signal can overload (desensitize) your receiver, causing received signals to appear weak. His loud signal may cause spurious signals to appear in the tuning range of your receiver. This is called intermodulation distortion (IMD). If you switch off your preamplifier or switch in the attenuator, you may be able to reduce the effective strength of his signal by an amount that will let your receiver function satisfactorily. These circuits reduce the signal level that reaches the receiver mixer. The mixer stage is where most of the trouble occurs when too strong a signal level is present.

You'll probably want to use your preamplifier, but not your attenuator, on 10, 12, 15, 17 and 20 meters most of the time. The preamplifier will improve the signal-to-noise ratio (S/N). A well-designed preamplifier has a low noise figure that makes weak signals easier to copy. If the incoming signal goes directly to the receiver mixer (preamp turned off), the inherent mixer noise (internally generated) may be too great to permit copying weak signals. Mixers are generally noisy when compared to low-noise RF amplifiers. At frequencies below about 18 MHz, the atmospheric and man-made noise picked up by your antenna is louder than that generated within the mixer, so the preamplifier is of little value. It amplifies the noise with the received signal, and this doesn't aid weak-signal copy. The increase in receiver gain then causes you to lower the receiver audio-gain level.

Noise Blanker

Your noise blanker (NB) is perhaps the least useful of the modern "bells and whistles" associated with receivers. New hams expect this magical circuit to reduce or remove all forms of man-made noise. Unfortunately, noise blankers aren't effective in reducing most types of QRN. Automotive ignition pulses can be blanked by a good NB, but most power-line noise and related electrical hash (appliances, light dimmers, etc) has too fast a repetition rate or too wide a pulse width for the NB to deal with.

The unfortunate fact is that a noise blanker can greatly degrade the receiver dynamic range (DR). Strong signals sound distorted and mushy when the NB is turned on. The NB can cause all kinds of spurious signals (caused by strong signals in the band) to appear in the tuning range of your receiver. NBs are often more trouble than they're worth.

Some receivers have a threshold control for the NB. The higher you set the blanking level (clockwise), the greater the receiver-performance degradation. Use the lowest (counter-clockwise) setting possible, consistent with noise reduction, when using your blanker. But more importantly, don't expect miracles when you activate your NB.

Using the Notch Filter

The NOTCH control is useful when operating CW or SSB. It's a tunable rejection filter that nulls or notches out unwanted single audio tones. You'll hear beat notes when another ham operates close to your frequency or when AM broadcast carriers are encountered, particularly on 7 MHz. It can happen when you're using the SSB mode and someone transmits a steady carrier close to your frequency. These high- or low-pitched notes can be annoying and are often caused by hams tuning their transmitters on the air without switching to a dummy antenna. They may even impair your ability to copy the signal you're listening to. When this occurs, activate your notch filter and adjust the control for minimum response from the interfering note. A well-designed notch filter can reduce the note by 30 to 40 dB. You may not eliminate the note completely when attempting to null it, but it will drop sufficiently in level to make operating more pleasant and effective. These unwanted beat notes are also known as "heterodynes."

S Meters

Signal-strength meters are the least essential devices we find on receivers. It's easy to become enraptured with these gadgets when giving out signal reports. We need to remember that they're only able to provide *relative* indications of signal strength. Few, if any, S meters yield the same reading for a given signal. If we were to connect your receiver and mine to the same antenna, then tune the receivers to the same signal, my meter might read S8 and yours could indicate 10 dB over S9. It depends entirely on the receiver circuit and the setting of the S-meter sensitivity control. In other words, don't believe what you see when observing your S meter.

S meters are useful when comparing the performance of antennas. An amateur may ask you to give him a signal report as he switches quickly from antenna "A" to antenna "B," for example. Your meter enables you to tell him which antenna provided the highest meter reading. You may use the same technique when comparing two or more antennas at your location. Observe the meter as you switch antennas. This gives you an indication of which of your antennas delivers the loudest received signal.

S meters are also helpful in comparing band conditions from day to day. You'll need to check band conditions against a known signal at a given distance. Specifically, observe your S meter while listening to an amateur you talk to frequently, or to WWV. Keep track of the signal strength of her station over several days or weeks. This will provide a fair indication of propagation conditions from day to day.

If you service your own equipment, you may use the S meter for a visual indication during receiver alignment. The various circuits may be peaked for maximum meter reading when a signal generator is connected to the antenna jack of the receiver.

Because of the inconsistency of S-meter readings, we frequently refer to these instruments as "guess meters." It's good to remember that the S-meter report you hand out over the air may inflate someone's ego (40 dB over S9!) or it may anger the other guy if you tell him his signal is a mere S3. You could be wrong in both instances because your meter may be inaccurate. There is, perhaps, no better means for giving signal reports than by ear. A report on SSB such as "Q5 and readability 9" is entirely adequate and acceptable. During CW operation an RST 579 is fine, if that's how you read the other station's signal.

Automatic Gain Control (AGC)

Your receiver may have a selector switch for AGC FAST, AGC SLOW and AGC OFF. The speed is related to how quickly the automatic gain control (AGC) circuit releases when an incoming signal ceases to transmit. AGC action can be observed as you watch your S meter. The needle of the meter will drop to the left quickly in the FAST mode. It will tend to hang and drop slowly in the SLOW mode. There will be no S-meter action if you turn the AGC switch to OFF.

Fast AGC is preferred by most CW operators, especially if they operate QSK. The receiver recovers more quickly when you stop transmitting. I prefer the slow-AGC mode for SSB operation. Slow AGC decay is also beneficial when the band is noisy or when weak signals are nearby in frequency. You don't hear the noise and interference between the words or code sent by the other station, unless there are large gaps between his words or code characters. Fast AGC permits these sounds to be heard between bits of information.

Unfortunately, slow AGC can be a deterrent if a strong signal is near your frequency. Its signal tends to "capture" the AGC circuit and desensitize the receiver. If the interfering signal is louder than the one you're trying to copy, you may miss many parts of his message. This can be minimized by using the fast AGC function. If you switch to the AGC OFF position, you can prevent your receiver AGC from "locking up" and causing desensitization when a strong signal is near to your operating frequency. When you disable the receiver AGC it's necessary to control the overall gain manually. This may be done by advancing the AUDIO GAIN to approximately mid range. The RF GAIN is now used to set the level of the received signal for comfortable volume. The S meters in many receivers become inoperative when the AGC is disabled. Don't be alarmed if the needle in your meter doesn't move as you tune in a signal or if the S-meter needle rises as you adjust the RF GAIN control.

Normal and Narrow SSB and CW Filters

Modern receivers have a switch that enables you to select wide or narrow IF filters. The manufacturer generally supplies the receiver or transceiver with "stock" filters. These are the so-called "wide" filters. A 2.4-kHz SSB bandwidth is typical, sometimes along with a 600-Hz bandwidth CW filter. You can operate successfully with these filters if severe QRM isn't present. Narrow-bandwidth filters (accessory options) help remove the QRM, as shown in Fig 2-3.

Narrow CW filters with a 250-Hz bandwidth are usually available from the manufacturer. For SSB reception, you may purchase a 1.8-kHz IF filter. The 1.8 and 2.4-kHz bandwidth filters compare to the illustrations in Fig 2-3. However, you should be prepared to accept poor audio fidelity when you use a 1.8-kHz SSB filter. The narrow bandwidth restricts the high- and low-frequency response of the voice you listen to. This makes the signal sound somewhat "tinny," but you'll still be able to understand what the other person is saying. Check with the manufacturer about the availability and cost for narrow-bandwidth filters that may be added to your receiver.

The QRM-reduction properties of narrow SSB filters aren't as dramatic as those of narrow CW filters. A 250-Hz CW

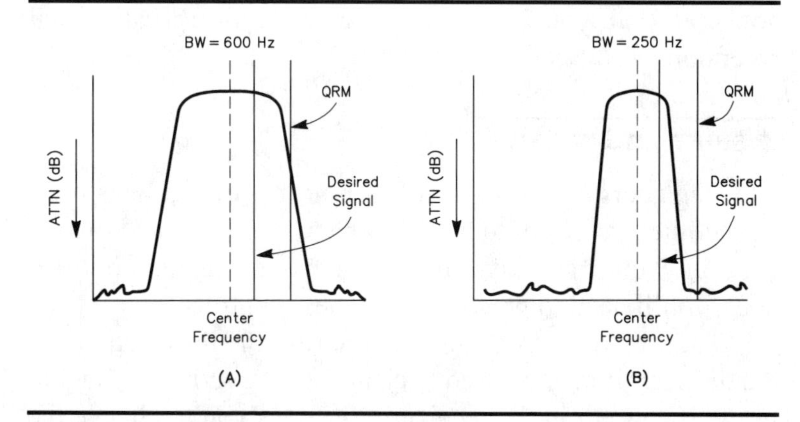

Fig 2-3—Simplified examples of 600 and 250-Hz IF filter band-pass characteristics. The curve at A shows how an interfering (QRM) signal can fall within the IF-filter response curve and be heard loudly with the wanted signal. The narrower filter at B allows reception of the desired signal, but the QRM-causing signal falls outside the filter passband, where it may be attenuated as much as 50-60 dB.

filter, for example, can improve the signal-to-noise ratio of your receiver during weak-signal copy. The filter elevates the incoming signal above the noise threshold. Consider for the moment that you're trying to copy a signal that's about the same strength as the atmospheric and man-made noise you're hearing. You're unable to enjoy solid copy because the signal is in the noise part of the time. Now you switch in your 250-Hz filter, then carefully readjust the main-tuning dial for maximum signal loudness or maximum S-meter reading. You'll notice that the weak signal is now louder than the noise and you have Q5 copy.

Your narrow CW filter will cause the signal to sound a bit "ringy" compared to a wider filter. Static crashes also have a ringing to them. This is a worthwhile trade-off when you consider that you've vastly improved your weak-signal reception.

IF SHIFT and WIDTH Controls

Beginners have difficulty learning how to use the IF SHIFT (sometimes referred to as "passband tuning") and WIDTH controls. Your transceiver may not have this feature, but if it does, you'll need to practice using these controls.

The WIDTH control is used to improve the IF selectivity of the receiver. It may be used to change the effective bandwidth of the IF system in the receiver. It enables you to change the IF bandwidth over a value that's set by your IF filter, then down to a very narrow IF bandwidth. It doesn't provide the same filtering quality a good IF filter can deliver, but it's useful when you have only the stock filters in your rig.

You can use your IF SHIFT to move the center frequency (Fig 2-3) of your IF circuit up and down (slightly) within the passband of the IF filter. This lets you shift an interfering signal outside the IF passband, which reduces interference from a nearby frequency. The IF SHIFT is moved in one direction from center for USB reception, whereas it must be turned the opposite direction for LSB reception. For CW reception, it's usually tuned in the same direction as for USB reception. Under normal conditions of reception, these controls are set at zero or midrange. Practice adjusting these controls until you learn their effects. Once you master them, it'll be second nature to adjust them quickly when they're needed.

RIT and XIT Controls

RIT stands for "receiver incremental tuning." Conversely, XIT signifies "transmitter incremental tuning." RIT is used when you want to maintain a given transmit frequency, but need to move the receiver tuning slightly (roughly ±1.5 kHz). This feature is useful when the person you're working isn't quite on your frequency. Adjust the RIT until his signal has a normal sound—especially during SSB reception.

XIT is used when you don't want to change the frequency to which your receiver is tuned, but do need to move your transmit frequency slightly. XIT isn't used as frequently as RIT. Some amateurs use XIT when working DX. The receiver stays on the DX station's frequency, but the XIT feature permits the DX chaser to call the DX station off his frequency (standard practice) to avoid the mayhem of a massive pileup (many stations calling him at the same time on the same frequency). The DX station often listens 1 or 2 kHz above or below his operating frequency to find a calling station in the clear. The RIT control is sometimes called the clarifier.

Receiver RF GAIN Control

Earlier in the chapter we discussed the RF GAIN control with relation to the AGC OFF function. Actually, most RF GAIN controls regulate the IF gain rather than that of the RF amplifier. Therefore, RF GAIN is, in a sense, a misnomer. You can use this control with the AGC operating. Under some strong-signal conditions, it will reduce potential receiver overloading because it reduces the signal level that reaches some stages in the receiver and this prevents these stages from being driven too hard (into nonlinearity). Your S meter won't function normally at reduced RF GAIN setting. The readings will differ from those obtained with the RF GAIN fully on.

Using a Power Amplifier

High-power (QRO) operation is an option provided by a linear amplifier. If you have a Novice license, there's no need to purchase a linear amplifier because your maximum power limit is 200 W. But if you hold a Technician or higher class of license, you may want to use an amplifier. An amplifier can change a weak signal into a Q5 one. For example, if your transceiver has 100 W of output power, a 1.5-kW peak envelope power (PEP) output amplifier will increase your signal strength by 11.76 dB. This is approximately two S units. Thus, if your signal is being received, say, S4, and the noise level at the other person's station is S4, your signal will rise to S6 with the amplifier in use. This elevates your signal two S units above the noise and it will be Q5.

Your power amplifier should be used as a tool—not a weapon! Never use it to "clear the frequency" or to drive away operators on a nearby frequency. This is defined as intentional interference and can earn you an FCC citation. It's clearly in violation of the Amateur's Code, which all hams should adhere to at all times. Using a linear amplifier for ragchewing across town or when working a station that hears you well without your amplifier is a violation of FCC rules. The rules state specifically that amateurs shall use no more power than is necessary to maintain communications.

The term "linear" indicates that the amplifier will reproduce and amplify the signal applied to the amplifier input circuit. It does this without creating distortion if you adjust the amplifier properly and don't apply too much driving power. This is necessary when you amplify SSB (and AM) signals. Linearity isn't required for the amplification of CW and FM signals. You can use a linear or a Class C amplifier for CW and FM operation. The amplifier linearity is established by providing a specific amount of operating bias (negative grid voltage for most power tubes, and positive voltage for power transistors). The value of bias depends on the tube or transistor used in the amplifier. This bias voltage causes a specified value of no-signal plate or collector current to flow. The resting or idling current sets the conduction angle of the tube or transistor performance. A Class C amplifier, on the other hand, has no resting current during no-signal periods.

Amplifiers Can Cause Problems

When you increase your transmitter power to the legal limit (1.5 kW), you increase the potential for RF interference to hi-fi gear, TV and radio broadcasting receivers, VCRs and telephones in your neighborhood. Furthermore, you often need a 220-V ac source in your ham shack for powering a big amplifier. Some amplifiers are designed to work from the 110-V ac line, but they can cause the household lights to flicker when you speak into the mike or key your transmitter. The picture on the family TV set may shrink and expand as you operate CW or SSB. These are problems that result from poor ac-line regulation when you place the typical 10-A or greater amplifier load on the line. Most linear amplifiers draw 20 A or more when delivering full rated power. In addition to the line-voltage changes, you'll observe that the amplifier plate voltage drops when you key the transmitter. This may be caused by poor line-voltage regulation. The peak output power of the amplifier is limited when this occurs and the amplifier efficiency could suffer.

Avoid these annoyances by operating your amplifier from a 220-V ac line. Ideally, a separate 220-V line should be brought to your shack from the main breaker box in your house. It's possible to bridge the 220-V line that feeds your electric range or electric clothes dryer, but you must avoid operating the appliances when the linear amplifier is in service. Too great a load will trip the circuit breaker for that line.

Fig 2-4 shows a simplified panel arrangement for a linear amplifier. Your amplifier may have only one meter, depending on the model. The switch labeled "C" permits the multimeter to indicate various operating conditions, such as tube plate voltage (I_E), current (I_P) and relative RF output power. Meter B may indicate only the tube grid current. In most grounded-grid amplifiers, it's important to keep track of the grid current to avoid exceeding the safe value for the tube or tubes in your amplifier. Excessive excitation from your exciter shows up as

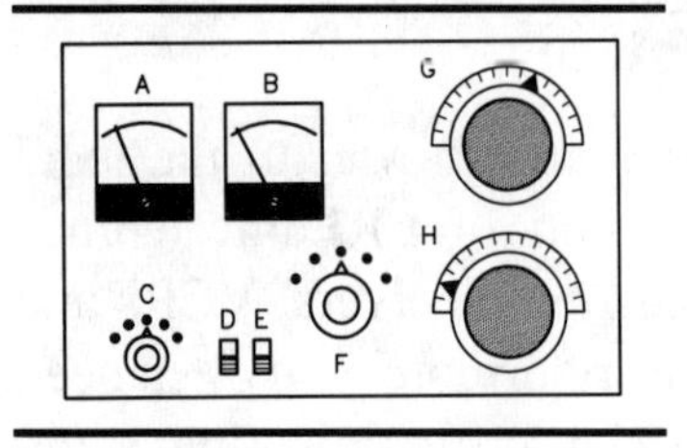

Fig 2-4—Panel layout of a typical HF linear amplifier. Multimeter A reads plate current (I_P and voltage (I_E). Meter B indicates grid current. Control C is the meter switch, D is the power-on switch, E is the standby switch and F selects the operating band. Control G is for plate tuning and control H is the loading capacitor. Some amplifiers have only one meter.

a high reading of grid current, and too much current will quickly ruin the tube. Check your operating manual to learn the safe maximum grid-current value. The same is true of the plate current. Too much excitation causes excessive plate current. This can cause tube damage. If you construct your amplifier from scratch, study the tube manufacturer's specification sheet for the tube you use. Don't exceed the maximum safe voltage and current values.

The tube life can be shortened if the filament voltage is too high or too low. Most commercial amplifiers are designed for the correct filament voltage, assuming that the ac line voltage is within normal limits.

Some amplifiers contain two or more tubes in a parallel hookup and other amplifiers have one tube. For example, some modestly priced amplifiers use four TV sweep tubes in parallel. Another model may have a single 3-500Z tube or two of these tubes in parallel. Some older designs call for two or four 572-B tubes in parallel, and many modern amplifiers contain a single power tube that can produce up to 1.5 kW of output power. When you purchase an amplifier, it's wise to learn the prices of various tubes. Replacing a tube can be as expensive as buying a used amplifier in good operating condition!

Generally, sweep tubes and 3-500Z tubes are the least expensive when it's time to replace a tube or tubes.

Amplifier Switches

Modern amplifiers have two main switches (D and E in Fig 2-4). One turns on the power, which applies voltage to the tube filaments and cooling fan. The other switch is for amplifier standby. When the STANDBY switch is in the OFF position, the exciter is routed around the amplifier for low-power ("barefoot") operation, but the amplifier is warmed up and ready to use when the need arises. To change from low to high power, you only need to adjust the transceiver's output power for proper amplifier excitation or drive, then set the STANDBY switch to OPERATE. The tube operating voltage is applied in the OPERATE mode. A changeover relay in the amplifier then handles the standby and operate function. This relay is controlled by a relay in the transceiver. A control line is connected between the transceiver and the amplifier.

ALC Circuit

The automatic limiting control (ALC) is used to prevent your amplifier from exceeding a specific peak-output power level. Most transmitters and transceivers have an ALC jack on the rear panel. A shielded two-conductor cable is routed from this jack to the ALC output jack on the amplifier. Check your operating booklets to learn how the ALC circuit should work with the units you own. The equipment may require an adjustment to ensure correct ALC operation.

The ALC circuit samples RF energy in the amplifier. This energy is fed back to the transmitter or transceiver to operate a sensing circuit. This sensing circuit reduces the transceiver output power (amplifier excitation power) when the amplifier power reaches a predetermined level. This helps to extend the amplifier tube life and it prevents you from accidentally operating with illegal output power.

You can operate your amplifier without using the ALC feature. If this is your choice, you'll need to keep a close watch on the amplifier grid drive and plate current to stay within safe and legal operating limits. Too much amplifier excitation reduces tube life and can cause your signal to be broad and distorted. Excessive bandwidth and spurious products are illegal and disrupt communications elsewhere in the band you're using.

Tuning Your Tube-Type Amplifier

The first step when tuning-up for a given frequency is to adjust your transmitter or transceiver in accordance with the manufacturer's instructions. Use a dummy antenna when doing this. You may now turn on your amplifier and allow at least a one-minute warm-up period. Connect the dummy antenna to the amplifier output jack. Activate your amplifier and apply a small amount of excitation from the transceiver. For example, if the maximum safe amplifier plate current is 400 mA at the dip, adjust the drive for a plate-current reading of 75 to 100 mA. This will permit you to go through the dip-and-load routine without harming the amplifier tubes. When the PLATE TUNING and LOADING is adjusted for maximum indicated RF output power, you may increase the excitation gradually until the rated plate and grid current is observed on the meters. Don't allow the amplifier to operate with a steady carrier for more than 10 to 15 seconds at one time or the tubes may become too hot. Allow a 30-second cooling period between adjustment periods.

You can attach your antenna after the tuning adjustments are finished. Apply a reduced amount of amplifier excitation and make sure the SWR is low (less than 2:1 is okay; this assumes you have an SWR indicator in the coaxial line between the antenna and the amplifier or between the amplifier and the antenna tuner) is low. Make sure you're on a clear frequency when you test with an antenna connected. Identify your station at the beginning of all tests. If you use an antenna tuner, make the preliminary tuner adjustments at low power with the amplifier off line. When the SWR is close to 1:1, you may increase the excitation for full power output. We'll discuss antenna-tuner adjustment in another chapter. Always check your amplifier-tuning adjustments for correctness after the antenna tuner is set for an SWR of 1:l. Changes in antenna-tuner values can affect the tuning of the amplifier, especially when the SWR is high.

You may find that your solid-state transceiver won't produce its rated output power when you connect it to an amplifier. This isn't uncommon. The cause of low output power from the transceiver is a high SWR at the input to the amplifier. When this condition prevails, the SWR-protection circuit in the transceiver limits the transceiver output power to protect the output transistors. The high SWR usually results from poor amplifier design or misadjusted amplifier input-matching networks. It's vital that the amplifier input circuit have 50-Ω characteristic impedance for the transceiver to mate with the amplifier. Transceivers that have tubes in the final stage will normally work okay when the amplifier input SWR is high. The transceiver tuning and loading controls may be adjusted to compensate for the mismatch. The practical way to cure the mismatch problem is to readjust the amplifier input-matching network for low SWR or you can install a small antenna tuner between the transceiver and the amplifier. The disadvantage of the latter approach is that the tuner needs to be adjusted each time you change bands or make large changes in operating frequency within a band. A small antenna tuner will accommodate 100 W or less of RF power.

Safety First!

Operator safety is the first order of business when working with amateur equipment. AC and dc voltages are always present and present lethal potentials in equipment that operates from the ac mains. RF voltages can burn flesh, so never touch coils, capacitors and antennas that are "live" (energized). Follow these rules when operating or repairing your radio equipment:

1) Connect the equipment cabinets to a quality earth ground.

2) Don't touch your antenna when transmitting.

3) Keep the radiating part of your antenna out of reach of people and animals.

4) Don't touch exposed terminals that have ac, dc or RF voltage on them.

5) Don't operate high-voltage equipment unless it's in a cabinet.

6) Keep your antennas a safe distance from power lines. Never erect an antenna or tower near enough to a power line to allow it to fall onto the power line.

7) Keep one hand in your pocket when measuring dangerous ac and dc voltages, and don't stand on a concrete floor or damp ground when measuring voltages.

8) Use a professional-quality safety belt when climbing your tower. Wear a hard hat when you work on antennas or towers.

9) Don't place a hot soldering iron on the workbench. Use a metal soldering-iron stand to prevent fires.

10) Some components, such as capacitors, can maintain a powerful charge even after you shut off the rig. Don't assume that it's safe to stick a finger in a chassis just because it's unplugged.

This list identifies the major hazards associated with Amateur Radio. Common sense dictates the general safety precautions to take when enjoying your radio pastime. Many experienced hams have lost their lives by ignoring these safety rules!

Equipment Malfunction

Aside from occasional circuit malfunctions caused by defective parts, you may have problems caused by unwanted RF energy getting into your station equipment. This is a common event that can baffle and confound even an experienced amateur. RF current sometimes follows unintended paths. When this happens, you may observe that the mike, key or equipment cabinet feels "hot" to the touch. The metal "bites" you when you touch it. This means RF energy is not only on the antenna, but it has remained in the ham shack to affect items that should be "cold" to the touch. One cause of this is an ineffective earth ground for the ham station. Instead of flowing to ground, the RF current seeks an alternative ground, which may be the ac line and the ineffective station ground (which includes the equipment cabinets).

When RF energy follows the wrong path, you might have circuit malfunctions, such as squealing and howling supermosed on the transmitter audio when you transmit. This unwanted RF energy can enter your CW keyer and cause it to send gibberish. Severe stray RF voltage and current can disrupt the transmitter and cause it to self-oscillate. It can also affect the transmitter's VFO stability. This disruptive energy is sometimes called "RF feedback."

Avoiding Unwanted RF Currents

A quality (low-resistance) earth ground is vital to smooth operation of your amateur station. Two or more 8-foot copper ground rods can be driven into the soil close to your radio room. The rods should be about 6 feet apart and bonded to each another with heavy copper strapping or braid. The connections are most effective if they're soldered to the rods with a propane torch or high-wattage soldering iron. Four such ground rods in a square are worth considering. The ground lead from the rods to the shack must be a heavy (large diameter or width) conductor to ensure an effective ground system. Long, thin connecting straps are inductive at RF. This presents an unwanted ac resistance that degrades the ground system. The larger the cross-sectional area of the conductor, the lower the inductance. You can use shield braid from RG-8 coaxial cable or strips of flashing copper for ground straps.

The effectiveness of the station ground may be enhanced by connecting it to the cold-water pipes in your house, assuming that you have copper plumbing with soldered joints. Chain-link fences may be added to the ground system. Bare no. 14 or 16 copper wire (several runs, as lengthy as possible) can be buried and copper plumbing with soldered joints can be put in the soil to improve the ground system. This helps when other measures don't cure the problem.

Copper straps or braided copper lines may be used to join the cabinets of your station equipment. This helps to minimize the adverse effects of stray RF energy in the radio room. Keep the ground leads as short as possible for best results. For example, you can use the ground straps to join the exciter cabinet to the case of the linear amplifier. Likewise for joining the antenna tuner cabinet to the exciter and amplifier. These straps are connected to the main ground line in your shack.

Earth Ground and the Station Location

Basements and ground-floor locations are best for keeping the station close to earth ground. You may experience difficulty with stray RF energy if you install your ham station on a second or third floor in a dwelling. This causes the earth-ground lead to be long and inductive. In fact, it may become resonant at some amateur frequencies. When this happens, the shack end of the ground wire may be hot with RF voltage. Some hams resolve this problem by using a second antenna tuner (low power) to tune out the reactance of the ground wire. You can build or buy a tuning device designed for this purpose (often called an "artificial ground"). A ground-lead tuner doesn't take the place of a proper earth-ground system; it helps remove unwanted RF energy from the ham station. A tuned ground lead presents an inconvenience to an operator because it's necessary to readjust the ground tuner whenever the operating frequency or band is changed. The situation is similar to readjusting an antenna tuner when changing frequency. Others lay a random length of wire around the shack to create a ground "counterpoise" to help cure RF feedback.

Some Antennas Cause RF Problems

Random-length end-fed antennas are troublemakers when the end of the antenna is routed directly to the ham shack. End-fed ½-wavelength antennas (or multiples of ½ wavelength) encourage the presence of unwanted RF energy in the station. As the operating frequency is increased, for instance, from 80 to 20 meters, the condition worsens, especially if there's a long ground lead between the radio gear and earth ground. The end-feed system brings high RF voltage into the radio room if the antenna feed impedance is high on a particular frequency. The end of a ½-wavelength wire has an impedance of several thousands of ohms, which accounts for the high RF voltage at that point on the wire. A random-length antenna may present a high impedance at the shack end of the wire. The lower the antenna-feed impedance, the lower the RF voltage, by virtue of $E = \sqrt{P \times R}$, where E is the RF voltage, P is the transmitter output power in watts and R is the feed impedance of the antenna in ohms.

Tuned feeders, such as 450-Ω ladder line or open-wire line, can cause RF voltage to appear in the ham station. This may occur when center feeding a multiband dipole or end-feeding a Zepp antenna. The feed line may reflect a high impedance back at the shack on some frequencies and this brings high RF voltage into the shack.

You'll have the least trouble if you feed your antennas with coaxial transmission line and maintain a low SWR (less than 2:1) in the overall antenna system. Coaxial-cable fed Yagis, cubical quads, verticals and ½-wave dipoles, when matched to the feed line, cause the fewest problems with stray RF energy inside the station.

How to Use Your SWR Meter

A useful, necessary station accessory is an SWR indicator or RF power meter. SWR indicators are known as SWR meters or SWR bridges. Not all SWR detectors contain the classic "bridge" type of circuit; it's sometimes called a "reflectometer" circuit. These matters aren't important, provided the instrument gives accurate readings and is designed for the characteristic impedance of your equipment and feed lines—normally 50 Ω. The instrument must have provisions for reading forward and reflected power. This enables you to adjust your antennas or antenna tuner for minimum reflected power. A full scale *forward* reading and a zero reading of *reflected* power indicates a perfect match, or a 1:1 SWR.

Fig 2-5 shows two styles of SWR/power indicators. Simpler units may not have the meter scales calibrated in watts. Such meters are useful only for SWR adjustments and the meter indication may only be relative concerning the actual SWR. In this case, adjust your antennas or tuner for a zero reflected reading on the meter.

The instrument at the right in Fig 2-5 is handy because the operator can observe the forward and reflected readings at the same time. This feature enables you to quickly detect any fault that may occur in the antenna system as you operate. If you

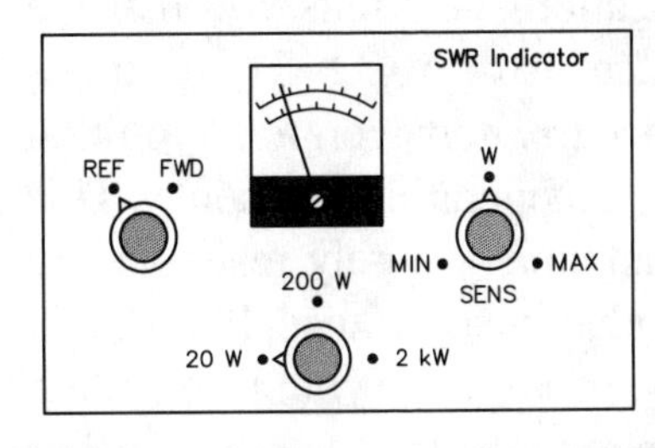

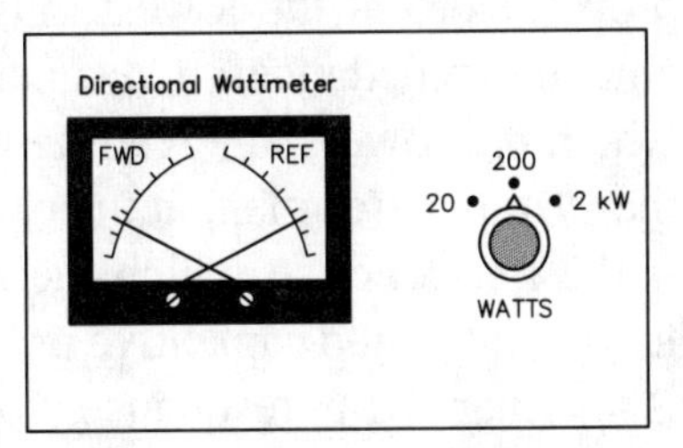

Fig 2-5—Two types of SWR indicators. The unit at the left has two meter scales. One is for RF watts and the other is an SWR scale. Only one meter is used, requiring that the function be switched from REF **to** FWD **when doing SWR measurements. The sensitivity** (SENS) **control at the right determines the meter deflection. It's adjusted to give a full-scale reading in the** FWD **mode when checking SWR. When the right-hand control is on watts** (W), **the meter reads transmitter-output power. The instrument portrayed at the right has crossed needles. One indicates reflected power and the other shows the forward power.**

were to monitor only the forward power with a single-needle meter, the SWR could suddenly become too high without your knowledge. This can damage the feed line, balun transformers or the output stage of a transmitter. Instruments are available with two meters on the panel; one reads reflected power and the other is for forward power.

Use your SWR or power meter with coaxial line that has the same impedance as the meter. The standard today is 50 Ω. If you use 75-Ω coaxial cable with a 50-Ω SWR meter, you'll obtain misleading, inaccurate readings because your instrument has been calibrated or balanced for a 50-Ω load.

Your SWR meter is an important tool for trimming dipole or vertical antennas to a resonant length. The antenna length is changed (power off!) and the result is noted on the REF scale of your meter. Trim the length equally on both sides of a dipole until you get the lowest reflected-power reading. It may not drop to zero at the lowest indication. This tells you that the antenna impedance, at resonance, isn't 50 Ω. The point at which the lowest REF reading occurs, however, tells you that the system is resonant at the chosen operating frequency. This is the frequency at which the antenna is purely resistive and has no unwanted capacitive or inductive reactance. If the SWR is less than 2:1 at resonance, you can operate your transmitter. Special matching procedures are required to lower the SWR to 1:1 if you can't obtain a zero REF reading on the meter. Your SWR meter may be used the same way when adjusting the matching section of a beam or other type of antenna equipped with a matching device at the feedpoint.

SWR meters are essential for ensuring that your antenna tuner is adjusted correctly. The tuner's control settings are varied as you view the REF scale of your SWR meter (installed between the antenna tuner and the transmitter or amplifier). Adjust the tuner's controls until the reflected power reading is zero. This reading lets you know that you've created a matched condition between the transmitter and the feed line. It doesn't mean that you've changed the antenna or feed line characteristics to correct a mismatch at the antenna feedpoint. The antenna tuner "disguises" the feed-point mismatch by adding or subtracting capacitive reactance or inductive reactance until the effective feed-point impedance is 50 Ω. Your transmitter will deliver maximum output power into the 50-Ω resistive load the tuner provides.

It may be necessary to readjust your antenna tuner to maintain a 1:1 SWR when rain, snow or ice gets on your antenna system. This is particularly true if you're using balanced feeders, such as 300-Ω TV receiving ribbon, ladder line or open-wire feeders.

The power reflected back to your transmitter (SWR other than 1:1) isn't radiated. For example, if you observe a forward reading of 500 W and a reflected indication of 100 W, subtract the 100 W from 500 W to obtain the 400 W of power being applied to your antenna's feed line. This set of numbers equates to an SWR of 2.6:1. The Transmission Lines chapter of the *ARRL Handbook* contains equations for converting power readings to SWR.

Some SWR indicators have a sensitivity (SENS) control (Fig 2-5). If you have this kind, adjust this control for minimum meter sensitivity (fully counterclockwise) before making SWR measurements. Place the FWD-REF switch in the FWD position. Key your transmitter and tune it for approximately ⅓ the normal output power. Now advance the SENS control until the SWR meter reads full scale (needle to the right). Initial antenna adjustment of tuning should always be done at reduced power, as indicated. This helps prevent damage to the transmitter when the SWR is high.

The next step is to switch to REF and observe the reading. If it's not zero, standing waves exist in the antenna system. If SWR is present, lengthen or shorten your antenna to reduce the SWR, assuming that the chosen operating frequency doesn't already coincide with the resonant frequency of your antenna. If the antenna is resonant at the operating frequency, but the SWR is high, you may need to do some matching between the antenna feedpoint and the transmission line. The length of the dipole or vertical antenna should be adjusted for the lowest SWR attainable. If your antenna has an adjustable matching section (such as with a Yagi beam antenna), adjust it until the SWR is 1:1. If you use an antenna tuner, adjust it until an SWR of 1:1 is obtained. Now you can increase the transmitter output power. Check between FWD and REF once more to ensure that a 1:1 match exists. Cut back on the SENS control setting when the power is increased. If the meter isn't at zero in the REF mode, do some final touching up of the matching system or antenna tuner.

If you use an RF power meter instead of an SWR meter, you may adjust your antenna, matching system or antenna tuner for minimum reflected power. These instruments don't always have an SWR scale in addition to a power (watts) scale.

Two important points to keep in mind: (1) Always install the SWR or power meter between the transmitter and the feed line or antenna tuner and (2) connect the SWR meter to the system with transmission line that has the same impedance as the SWR or power meter. If you try to use your SWR instrument in any other manner, you'll obtain false readings.

You may experience a situation in which the SWR can't be brought to 1:1. This pertains to "real" antennas and not dummy antennas. Even though a proper match is effected between the antenna and the feed line, the SWR reading might be, say, 1.3:1. How can this happen? It's not uncommon to find a residual SWR reading when the transmitter is rich in harmonic output (mistuned or faulty design). The harmonic energy is rejected by the antenna because the match is incorrect at that frequency and it shows up on the meter as an SWR other than 1:1. This can happen even though the SWR is 1:1 at the desired operating frequency. If you have difficulty obtaining a 1:1 SWR, check the transmitter tuning and be sure that the output stage isn't being driven too hard. This false SWR reading won't be observed when operating your transmitter into a dummy antenna. This is because the harmonics will be accepted by the 50-Ω resistive load.

Setting up a resonant antenna system for the frequency you're using and getting a 50-Ω match at the transmitter is important, but don't make the common mistake of being paralyzed by not being able to adjust everything for a 1:1 SWR. If your transmitter can handle it without damage, the actual SWR reading isn't something to be obsessive about. Get it as low as you can, by making the proper adjustments to the antenna, feed line and tuner, and if it's within your radio's safe operating range, go ahead and get on the air. Some hams seem to spend their lives trying to get a perfect 1:1 SWR from their systems, rather than operating their radios at all.

How to Use an Antenna Tuner

Let's define the term *antenna tuner* (sometimes called a matcher, matchbox, tuner, antenna tuning unit ["ATU"] or Transmatch). It's a network of coils and capacitors configured to provide an impedance match between the transmitter and the antenna feed line. If the tuner is between the transmission line and the antenna feedpoint, it's capable of matching the antenna to the feed line.

Various hookups are used in antenna tuners. The coil and capacitors may be arranged to provide the more popular T network, such as that popularized by Lew McCoy, W1ICP, in his July 1970 *QST* article, "The Ultimate Transmatch." Other circuits are set up as pi (π) networks, parallel-resonant tuned circuits with coil taps for the feeders, or as the SPC Transmatch. It doesn't matter which circuit you use, provided the matcher can be made to establish an SWR of 1:1.

The coil inductance and the capacitances are adjusted to cancel (tune out) inductive or capacitive reactance at the transmitter end of the feed line. When these reactances are canceled to provide a 50-Ω resistive load for the transmitter, you have a 1:1 SWR.

You may have problems with RF voltage arcing between the tuning-capacitor plates when using your antenna tuner with high antenna impedances. Sometimes the feed line presents these impedances to a tuner. It's not uncommon to get annoying arcing in parts of the tuner when this happens. The ailment becomes more pronounced as the transmitter output power is increased. You can avoid this by matching the feed line to the antenna. This ensures that the transmitter end of the feed line presents a 50-Ω impedance, or nearly so, at and near the frequency of antenna resonance. Tuning capacitors with wide plate spacing minimize capacitor arcing. Spacing of ⅛ inch or greater between the rotor and stator plates is best for preventing arcing of the tuning capacitors.

Although you may obtain a proper impedance match between the feeder and the antenna feedpoint, there can be a mismatch elsewhere on the band. For example, if you adjust your 80-meter dipole for an SWR of 1:1 at 3700 kHz, the SWR can rise to 3:1 at 3600 or 3900 kHz. This is because antennas have a particular bandwidth. Antenna bandwidth is arbitrarily defined as the range of frequencies where the SWR is 2:1 or less. The quality factor (Q) of the antenna versus operating frequency determines the antenna bandwidth. The greater the antenna Q, the more narrow the frequency response. Small-diameter antenna wire increases the Q. Greater bandwidth can be obtained by using large conductors or several small conductors spaced apart and in parallel. For example, no. 10 copper wire provides greater bandwidth than no. 22 wire. The lower the operating frequency, the narrower the antenna bandwidth. You may observe an SWR no greater than 2:1 across all of the 15-meter band when using a wire dipole. On the other hand, you may have only 100 kHz of SWR bandwidth (less than 2:1) with a wire dipole on 80 meters. (As VHF/UHF operators know, a rubber duckie or a vertical ¼-wavelength whip can have an SWR of less than 1.5:1 from 144 to 148 MHz—a 4-MHz bandwidth!)

This is where your antenna tuner comes in handy. If your HF dipole has a high SWR at some frequency, you can use your antenna tuner to offset the antenna mismatch. The advantage is that by providing a 50-Ω load for the transmitter, it will deliver its full rated output because the internal SWR protection circuit won't automatically reduce the power.

Fig 2-6 shows the panel layout of a typical antenna tuner. The example shows a built-in SWR indicator. If your tuner doesn't have this feature, you can install an external SWR meter.

Initial antenna tuner adjustment should be done at reduced power for the reasons mentioned earlier. Pick a clear frequency before tuning up. Identify your station, then proceed with the

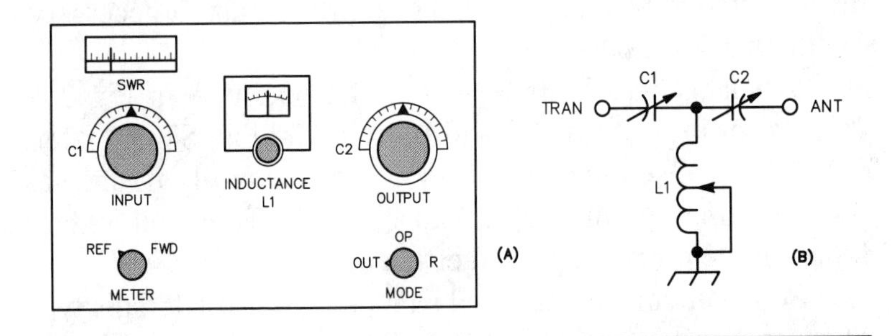

Fig 2-6—The panel of a typical antenna tuner (A) and the circuit for a T network (B). The SWR detector and meter is used at the tuner input, immediately after the transmitter. The MODE **switch permits bypassing the tuner, to use it or to route the transmitter output power to a dummy antenna (R).**

adjustments. The SWR may be too high at the beginning to allow transmitter use, but identify your station as soon as the SWR is low enough to operate safely.

Begin your adjustments by listening on the chosen operating frequency. Coarse adjustment of the antenna tuner calls for putting the three controls at midrange. Adjust them alternately for the loudest receiver background noise. This procedure helps you find control settings reasonably close to the final ones when the transmitter is operating.

Apply reduced transmitter power and observe the reflected-power reading on the SWR meter. Adjust each antenna tuner control for the lowest SWR reading. The controls interact with one another, so it's usually necessary to touch them up two or three times until the SWR is 1:1. Log the control settings for future use. This will help you change frequencies and tune up more quickly than if you hunt for the correct settings each time you change bands. You may now increase the transmitter power to the normal operating level. Recheck the reflected power and adjust the controls again, if necessary, to obtain a 1:1 SWR.

When using a T network like the one at A in Fig 2-6, various settings of C1, C2 and L1 will yield an SWR of 1:1. Select control settings that allow C2 to be as fully meshed as possible (maximum usable capacitance). This will reduce losses in the tuner (improved efficiency).

Some antenna tuners used a tapped coil for L1. This type of tuner can be restrictive, depending on the antenna system impedance, and it may not be possible to reduce the SWR to 1:1; it may fall to only about 1.5:1. A roller coil, such as that shown at B of Fig 2-6, provides a wider matching range because ¼ or ½ turn of coil may be needed to obtain an SWR of 1:1.

Antenna Tuners with Balun Transformers

Perhaps your antenna tuner contains a balun transformer. If so, it can be used when the antenna is equipped with balanced feed line, such as TV twinlead, ladder line or open-wire feed line. This transformer is a broadband device that should work from 1.8 to 30 MHz. It may have a 1:1 or 4:1 impedance-transformation ratio. It converts the unbalanced 50-Ω coaxial line from your transmitter to a balanced line of a higher impedance. It seldom functions in this application as an impedance-matching device because the balanced feed line may reflect several impedances to the tuner. This is especially true when you use balanced feed line with a multiband dipole antenna. The balun transformer can become hot and arc internally when high RF power is used to feed a balanced line that reflects a high impedance (greater than 500 Ω). A balun transformer is operating in the worst possible environment when used with a multiband dipole antenna. The exception is when a trap dipole is used and the system presents a low impedance to the balun transformer at all times.

Outboard Balun Transformers

Some amateurs install balun transformers at the antenna feedpoint to prevent the antenna balance from being disturbed when using coaxial feed line. There's little gained by doing this on the HF bands, especially when using dipole antennas—a balun transformer used like this can cause losses in the system and increase the SWR. All passive devices, such as transformers, by their nature introduce power loss in the system. The losses increase when the balun is used in a mismatched environment. An example of this is when you attempt to use a balun at the feedpoint of a multiband, center-fed Zepp dipole. Overheating and arcing can occur, just as it can in an antenna tuner balun.

Don't use balun transformers in an antenna system that doesn't have well-defined performance characteristics. If you have a commercially made antenna, check with the manufacurer before using a balun with it.

Glossary

AGC—Receiver automatic gain control. Sometimes called automatic volume control (AVC). An internal circuit that keeps the receiver audio output at a relatively constant, preset level, despite large changes in incoming RF signal level.

ALC—Automatic level or limiting control. Helps to prevent a linear amplifier from exceeding a given power-output level. A dc voltage is developed from the amplifier RF output energy and fed back to a control circuit in the transceiver. The developed dc voltage reduces the driving power to the last transmitter stage in the transceiver, thereby limiting the transceiver driving power to the amplifier.

Antenna tuner—A network of coils and capacitors that permits matching the transmitter output port to a feed line or antenna. Usually called a tuner, antenna tuning unit (ATU), Transmatch or coupler.

ANTIVOX—An adjustable circuit in a transmitter that enables the operator to use a loudspeaker during VOX operation. The ANTIVOX (or antitrip) circuit prevents the VOX circuit from activating in the presence of audio from the speaker.

Attenuator—A resistive 50-Ω network in a receiver's front-end circuit. When the attenuator is in use, it lowers the incoming signal by a set amount, such as 10 or 20 dB. This helps overcome receiver overloading when a strong signal is present.

Balun—*Bal*anced-to-*un*balanced transformer. A broadband transformer used to mate a balanced feedpoint or transmission line to any type of unbalanced coaxial line. A balun may provide an impedance transformation (for example, 4:1).

Drive—Driving power or RF excitation. The signal energy used to excite an amplifier. Drive is expressed in mW or W.

Earth ground—A physical connection to earth that has the least ac and dc resistance practicable. Usually accomplished by driving several long metal rods into the earth and bonding them to one another.

FWD—Forward power as indicated on an SWR meter or RF power meter.

IC—RF amplifier cathode current (tube) or collector current (transistor). This current is displayed on a panel meter.

IF SHIFT—A receiver circuit that has a panel control that enables the operator to move the receiver signal around within the receiver intermediate-frequency (IF) passband to reduce interference (QRM).

IF WIDTH—A receiver circuit whose front-panel control permits the receiver operator to widen or narrow the IF bandwidth.

Impedance —The ac resistance and reactance of a circuit, feed line or antenna, such as 50-Ω coaxial cable or 300-Ω TV ribbon.

Linear amplifier—An RF power amplifier used to increase a station's output power. A linear amplifier faithfully reproduces the waveform of the RF energy that drives the amplifier when it's adjusted correctly and not driven to excess.

Loading— A vacuum-tube transmitter control that's adjusted to provide the desired coupling between the transmitter and the antenna system. This occurs when the output tube is presented with the correct value of load resistance at its plate.

MOX—Manually operated transmit; operation of the transceiver changeover relay (transmit-to-receive mode change).

Noise blanker—A receiver circuit that may be activated to reduce or eliminate certain types of pulse noise that interfere with reception. Doesn't work on all types of noise, static or QRN, particularly that with wide pulse widths, such as static crashes from thunderstorms.

Preamplifier —Generally applies to the RF amplifier stage between the antenna and the mixer in a receiver.

PTT—Push-to-talk circuit. A microphone switch used to change from receive to transmit mode (for MOX or Manual Operation).

REF—Reflected power, as indicated by a wattmeter or SWR indicator in the antenna system. The opposite of FWD.

RIT—Receiver incremental tuning on a transceiver. Permits the operator to change the receiver frequency slightly without disturbing the transmit frequency.

SWR—Standing wave ratio between two RF devices or between a transmitter and an antenna.

Transmatch —See "Antenna tuner."

UHF—Ultra high frequency, generally applied to the bands between 300 and 3000 MHz. Frequencies above 1000 MHz (1 GHz) are usually known as microwaves and have specialized antenna and feed-line requirements.

VHF—Very high frequency, generally applied to the bands between 30 and 300 MHz.

VOX—Voice-operated transmitter relay. Used instead of MOX to control transmit/receive changeover by speaking into the microphone.

XIT—Transmitter incremental tuning. Allows the operator to change the transmit frequency slightly without disturbing the receiver frequency (opposite of RIT).

Chapter 3

Building and Using Antennas

W*hat should I use for my first antenna?* That's a good question. You may wonder, "How high above the ground must my antenna be?" These questions are important if you want good performance and to communicate over great distances.

Experimenting with antennas will help you determine which antenna works best to achieve your objectives. A particular type of antenna may work well at one location, but at some other location, the same antenna, erected at an identical height, may be a dismal performer.

Other factors include the antenna height above average terrain (HAAT) and the amount of conductive clutter at a given site. Clutter, such as steel-frame buildings, power lines, trees and phone lines, can distort the antenna pattern and absorb signal energy. Regarding antennas for HF in particular, much depends on the ground conductivity, which varies markedly throughout the US. Try various types of antennas to learn which one works best for you. Start with simple antennas to minimize cost and time.

Locating Your Antenna

Place your antenna as far from metal objects as possible. Keep the radiating part of the antenna away from your house. Your home's electrical wiring and plumbing lines are conductive objects that can affect antenna performance. Some houses have aluminum siding that may degrade the performance of antennas.

Don't be discouraged as you consider the foregoing points. There are few ideal antenna sites, especially for those who live in urban neighborhoods. Do the best you can and you'll enjoy good communications, even if your property is laced with power lines and such. Avoid erecting your antennas parallel to phone and power lines to minimize unwanted coupling to those lines. If you can, erect your antennas so they're at a right angle to power lines to minimize the bad effects of the antennas being near them. The right-angle installation will reduce unwanted QRN (pickup of man-made noise) and aid reception, while reducing unwanted absorption of RF energy during transmissions. ***Caution:*** **Do not erect your antennas so the wires are directly above or below the power lines. An antenna that comes in contact with a power line can be *lethal*!**

How High is "High"?

New hams tend to think of antenna height in terms of feet or meters. That is, 60 feet seems high as we look upward at a tower or treetop. A height of 30 feet may seem adequate for erecting a dipole or an end-fed wire antenna. These heights may be adequate for some frequencies, such as 28 MHz and higher, but at the lower end of the amateur frequency spectrum (160, 80 and 40 meters) these low heights don't permit spectacular antenna performance. For VHF and UHF, a high, clear spot is the key. Most signals above 50 MHz routinely travel only a bit beyond the line of sight to the horizon, so a low antenna is much less effective than one raised high atop a house, tree or tower.

HF antenna performance is determined to a large degree by the height above effective ground in terms of wavelength, rather than feet or meters. To be specific, a horizontally polarized antenna (that is, one whose main axis is parallel to the earth's surface) performs best when it's ½-wavelength or greater above ground. This dimension (in feet) is based on the operating frequency. Antennas that are close to ground aren't effective for long-distance communication and may lose their directional characteristics (become essentially omnidirectional). High-angle radiation occurs, and this isn't suitable for DX work. Table 1 indicates the height in feet for HF antennas that are ½-wavelength high. The heights are derived from the equation H = 492/f, where height is in feet and frequency is in MHz.

The table shows how the height above ground differs in terms of the operating frequency. A height of 60 feet, for

Table 1
Minimum Height Above Ground for Best HF Antenna Performance

Band	*Frequency MHz*	*½-Wavelength Height (ft)*
80	3.7	133
40	7.125	69
15	21.150	23
10	28.3	17½

example, is no height at all for 80-meter operation with a horizontal antenna. At the same time, we can appreciate the impracticality of having a 133-foot-high support for an 80-meter antenna. Operation at that frequency is generally a compromise for most amateurs.

What options do you have for, say, 80 and 40 meters? You must accept low antenna height and hope for acceptable performance. Your antenna should be erected as high as possible within your means and property limitations.

An alternative to low antenna height with horizontal antennas is the vertical antenna. Low-angle radiation (for DXing) and omnidirectional radiation are possible with vertical antennas mounted at ground level. A full-size vertical antenna is ¼-wavelength tall, but shorter verticals can be built. They require loading coils and capacitance hats to make up for the missing conductor length. Shortened verticals don't perform like full-size ones, but they're generally more effective than dipoles that are close to ground. Most man-made noise is vertically polarized, however, so you may hear more QRN when using a vertical antenna. We'll consider the vertical antenna in greater detail later in this chapter.

A Closer Look at Antenna-Height Effects

Too low an antenna height can inhibit DX performance. What can we expect from antennas less than ½-wavelength high? The lower the height, in terms of wavelength, the worse the performance. Consider, for example, a 3.7-MHz dipole only 30 feet above ground. It will show little directivity and most of the radiated energy will be straight up (very high angle). This is a benefit for short-range communication. This type of antenna may outperform a dipole at 133 feet (½ wavelength) for QSOs out to approximately 600 miles, but beyond that distance the signal falls off rapidly. Antennas like this are often thought of as "cloud warmers." Some amateurs erect them purposely for close-in work. The dipole that is at least ½-wavelength will, by comparison, produce excellent DX results, while being mediocre for contacts out to 500-600 miles. If you could see the energy being radiated from a low dipole, it would look like a sphere, whereas a high dipole has a bidirectional, figure-8 pattern off the broad side of the antenna. Construction details for cloud-warmer antennas are in *W1FB's Novice Antenna Notebook*.

Antenna height affects the feed-point impedance. A dipole ½-wavelength high exhibits an impedance of 70 Ω. The same antenna at 0.12 wavelength above ground presents an impedance of roughly 35 Ω. The feed impedance approaches 100 Ω at approximately 0.33 wavelength above ground. Therefore, it's incorrect to assume that all dipoles have a 70-75 Ω feed impedance. This explains why some dipoles can be fed with 50-Ω feed line and exhibit an SWR of 1:1, while others have a 1:1 SWR at 75 Ω. It depends on the antenna height. From a practical point of view, dipoles will perform well when the SWR is less than 2:1, so it's not imperative to reduce the SWR to 1:1 if your transmitter can accommodate the reflected power without its internal SWR protection circuit greatly reducing the transmitter output power. Many modern solid-state transmitters are equipped with automatic antenna tuners that will present the transmitter with a low SWR.

Radio-Frequency Ground

Radio ground isn't necessarily present exactly at the earth's surface. Significant RF currents can flow many feet below the surface of the earth. The conductivity and the dielectric constant of the earth determine how a radio signal interacts with the RF ground. These factors depend on the mineral and moisture content of the soil, which varies significantly from one geographic region to another. If you live on dry, sandy or rocky ground, the earth ground will be relatively poor compared to the situation if you reside near salt water or in a swampy area. Here the good RF ground may aid the transmission and reception of low-angle DX signals, particularly for vertically polarized antennas. Horizontally polarized antennas are also helped to a smaller degree over good RF ground.

Drawings that illustrate the patterns of horizontal antennas and vertical antennas over average ground compared to "perfect earth" are presented in the *ARRL Antenna Book.*

Artificial Grounds

Most amateur vertically polarized antennas depend on an effective ground system to perform well. Antennas ¼-wavelength long or odd multiples thereof require a good ground system. You may think of a ¼-wavelength antenna as half of a dipole. The feed line requires two terminals for connection

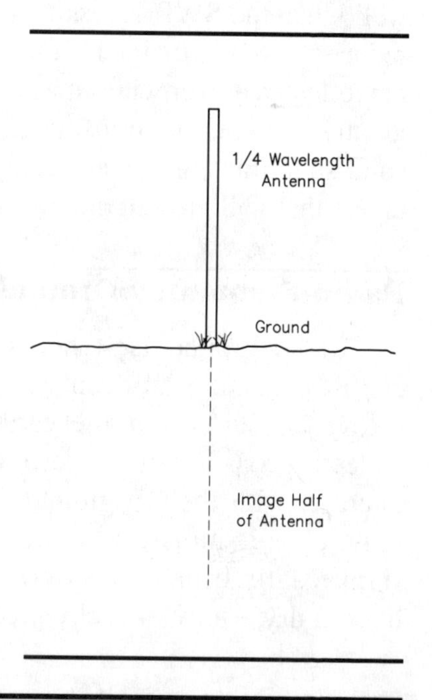

Fig 3-1—A ¼-wavelength vertical antenna. The dashed lines show the missing half or mirror-image portion of the antenna system. The radiator and its image can be thought of as a ½-wavelength dipole antenna. The feed line connects between the bottom of the radiator and the ground system beneath the antenna. Normally the artificial ground consists of an array of ¼-wave radial wires that lie atop the ground or are buried a few inches below the surface of the earth.

to the antenna, and the ground system provides the second terminal for a ¼-wavelength antenna. The missing half of the dipole exists as the "image" half in our ground system. You can't see or touch the image portion of the antenna, but it exists in your ground system as an invisible conductor. Fig 3-1 illustrates this principle.

No matter how conductive the earth may be at your location, there's no way for you to make a connection to it unless you provide a physical ground system that works with the earth ground. This can be realized in an ideal situation (100% ground conductivity) by driving a long metal rod into the earth near the antenna feedpoint. Normally, an array of 8-foot metal rods is driven into the ground and bonded together by a heavy conductor to ensure an effective ground system. I haven't seen an installation of this type that provided a truly effective *radio* ground, even though it can serve as a good dc ground for the station and may help protect you from lightning strokes.

Some commercial products sold as "artificial grounds" don't take the place of a true earth ground. They're used solely to tune out the reactance of a poor ground system or ground lead so that unwanted RF energy doesn't appear on or in the station equipment and disrupt the circuit functions (or "bite" you when you touch the gear while transmitting).

A man-made ground system is required to provide proper antenna performance. This is called a "ground screen" or "counterpoise" ground. The ground screen consists of numerous wire conductors (radials) laid on the ground or buried, after they are fanned out from the base of the vertical radiator. These wires are joined at the base of the antenna. The center conductor of our coaxial feed line is connected to the vertical element or its impedance-matching network. The shield braid of the feed line attaches to the common point of the radial system.

The radial wires can be elevated above ground on poles instead of being buried or laid on top of the ground. Above-ground radials need only be a few inches above ground if people and animals don't move about near the antenna, but they're usually installed 8 feet or more above ground so that people may walk safely under them. Good performance is reported when using only four such above-ground radials, whereas on-ground or in-ground systems require many radial wires. These on-ground wires, ideally, are ¼ wavelength or greater overall, but shorter wires can provide acceptable performance when space isn't available for long pieces of wire. You may use 8 or 16 radial wires if there's space for ¼-wavelength ground wires. As many as 120 radials may be used to ensure top performance in an elaborate installation. Chicken wire or metal fence wire may be used for developing a ground screen. Don't avoid using a vertical antenna if you lack the space for a perfect ground screen. Use as many radial wires as you can manage and don't worry if some of them are fairly short. You may even route them around your house or other nearby obstructions. They need not be deployed in a perfect 360° circle around the base of your antenna. Even though a poor ground screen causes increased system losses (reduced efficiency), your signal will still be heard at great distances. The ham in your area who has the same type of antenna and comparable transmitter power may be heard louder than you if he has a larger radial system, but you'll still be able to work plenty of DX with your smaller number of radial wires.

Your radial wires need not be of large diameter. Wire gauges from no. 8 to no. 22 are satisfactory, except for the fragility of the higher-gauge wires. A thin wire breaks easily and it will be destroyed more rapidly than the heavier wire when soil chemicals eat into the conductors. Most soil has an alkalinity or acidity characteristic and this causes corrosion of metal objects. Enamel-coated wire lasts longer in the ground

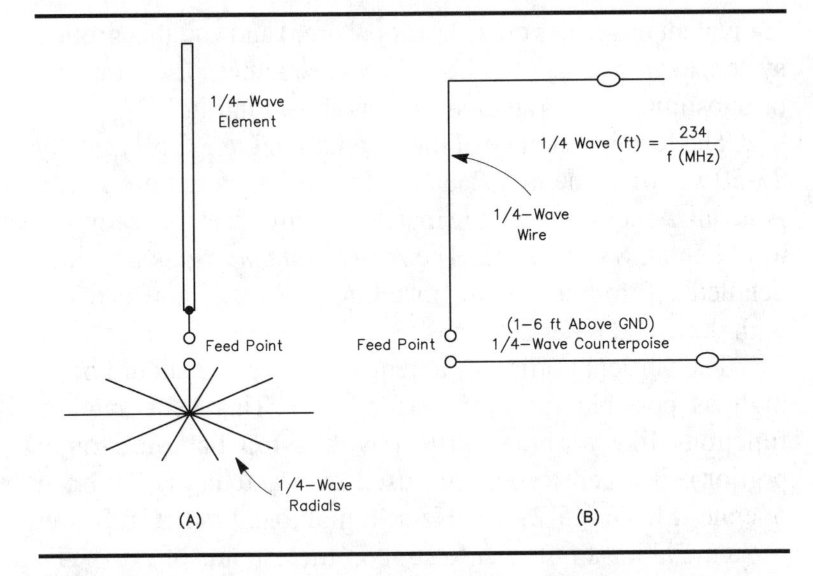

Fig 3-2—Antenna A is a conventional ¼-wavelength ground-mounted vertical with buried or on-ground radial wires (see text). Compromise antenna B has a bent ¼-wavelength wire element worked against a single ¼-wavelength counterpoise wire. Additional above-ground counterpoise wires will improve the antenna performance. The vertical portion of the radiator should be as tall as practicable for best results.

than bare copper or aluminum wire. I've found that no. 14 solid conductor, vinyl-plastic insulated electrical wire lasts many years in the ground. The plastic insulation protects the copper from soil chemicals.

The effectiveness of your ground screen can be enhanced if you connect other ground conductors to the radial system. For example, bond your chain-link fence to the ground system. Likewise with the cold-water pipes in your home and well pipes, if you have them. Improvement may be realized if you drive four 8-foot ground rods (4 feet apart) into the earth near the feedpoint of your vertical antenna. Join the rods with the shield braid from RG-8 coaxial cable (or a conductor of equivalent or greater cross-sectional area) and add this ground system to your array of radials. A single counterpoise wire may be substituted for a radial system as shown in Fig 3-2.

The feed impedance of the antennas in Fig 3-2 is typically 25-50 Ω, with antenna B exhibiting the highest impedance. Antenna A requires a simple matching network at the feedpoint to obtain an SWR of 1:1. The *ARRL Antenna Book* provides detailed information about matching networks you can use with these and other antennas.

The vertical portion of antenna B in Fig 3-2 should be as high as possible for best performance. This bent antenna functions like a short vertical with a top hat (horizontal portion). If tuned feeders are used with antenna B, you can operate it from 3.5-29.7 MHz as a multiband radiator. If you use a metal mast or tower to support the antenna of Fig 3-2B, keep at least two feet between the metal support and the vertical wire. The greater the spacing, the better. A tree can be used to support this antenna, but bare wire shouldn't come in contact with leaves and branches. Antenna B is commonly called an "inverted L."

The conductor size for vertical antennas isn't critical. You may use a tower or a piece of wire as the radiating element. The greater the cross-sectional area of the radiator, the greater the antenna bandwidth between the 2:1 SWR limits. This is because the larger the conductor size, the lower the antenna Q; and the lower the Q, the greater the effective bandwidth for RF circuits.

Choosing Your Feed Line

Coaxial (unbalanced) cable is the most common feeder for single-band dipole or vertical antennas. Tuned, balanced feed line is often used for multiband dipoles. Single-band dipoles may be fed with 50-Ω or 75-Ω coaxial line. If your transmitter power is less than 300 W or so, you can use RG-58 (50 Ω) or

RG-59 (75 Ω) coaxial line for feeding HF dipoles and other low-impedance antennas, such as verticals and beam antennas. At power levels above 300 W or at higher frequencies, use RG-8 (50 Ω) or RG-11 (75 Ω) cable. This larger cable can accommodate the higher power without heating or arcing, which happens if the SWR is high while using the smaller coaxial cable. Larger coaxial line has a lower loss per 100 feet than small-diameter line. Your overall antenna system will weigh more with heavier coaxial feed line. The weight of the cable can make a dipole sag in the center, whereas it may not do that with RG-58 or RG-59 line.

RG-8X coaxial cable is about ¼ inch in diameter, but it can accommodate greater power than will RG-58. I've used RG-8X at full legal amateur power and it hasn't failed, even when the SWR wasn't quite 1:1. It's lightweight and flexible. RG-8X is, however, more lossy (signal attenuation in dB) than RG-8 at the higher frequencies. Avoid RG-8X or RG-58 for VHF and UHF installations.

Be careful when routing any foam-filled coaxial cable, such as RG-8X, into your ham shack. You must make gentle bends to avoid pinching between the inner and outer conductors of the feed line. The soft inner insulation can compress and crush the inner and outer conductors close to each other. This encourages voltage breakdown and disturbs the impedance of the line by creating impedances other than 50 Ω at the pinched areas.

All feed line exhibits losses. Coaxial cable is rated in dB loss per hundred feet, depending on the operating frequency. The line loss at 2.0 MHz, for example, may be only a fraction of a decibel per 100 feet. For RG-58 it's 0.6 dB per 100 feet at 2.0 MHz. The same cable will cause a signal loss of 2 dB at 28 MHz and 4.5 dB at 144 MHz. This demonstrates the relationship of frequency to dB loss. Now consider dB vs RF watts: If we lose 3 dB in a length of feed line, we've wasted ½ of the transmitter output power. If your transmitter puts out

100 W, there will be only 50 W reaching the antenna. A 3-dB signal loss equates to ½ an S unit. This same loss will occur when a received signal comes through the coaxial cable. Consider what happens when we substitute RG-8 cable for RG-58. The loss per 100 feet at 2 MHz for the bigger cable is 0.22 dB, 0.9 dB at 28 MHz and 2.0 dB at 144 MHz. The larger cable is a better choice, irrespective of the power output of your transmitter. The *ARRL Antenna Book* has a table that shows the loss per 100 feet for most types of coaxial cable used by amateurs. Consult this book before selecting your transmission line.

VHF and UHF operation puts great demand on the capabilities of feed line. Above 100 MHz, feed line becomes a significant factor in determining how well your station will perform. For example, a 100-foot run of RG-58 coaxial cable will stifle more than half of your power output on the 420-450 MHz band. At higher frequencies, only a fraction of your transmitted and received signals will find their way through typical coax. Although it's more expensive and unwieldy, Hardline is the correct choice for UHF and above. For example, you'll find Hardline in ½- and ⅞-inch diameters, and at 200 MHz these have a loss of only 1.2 and 0.8 dB per 100 feet, respectively. It's more expensive than coax, at about $2.50 and $6 per foot, respectively, and a proper installation requires special connectors that can cost $30-$80, or more, each. It can cost $400, or more, for a 50-foot run of ⅞-inch Hardline with connectors. If you want to operate a serious UHF station, however, it's worthwhile to invest in this type of feed line. The initial investment may appear high, but it's not cost-effective to spend hundreds or thousands of dollars on transceivers, antennas, rotators, amplifiers, preamps, tower and other associated equipment, and then waste it all by letting the signals dissipate in the feed line.

If you plan to operate weak-signal modes, such as terrestrial SSB and CW DXing, collecting grid squares,

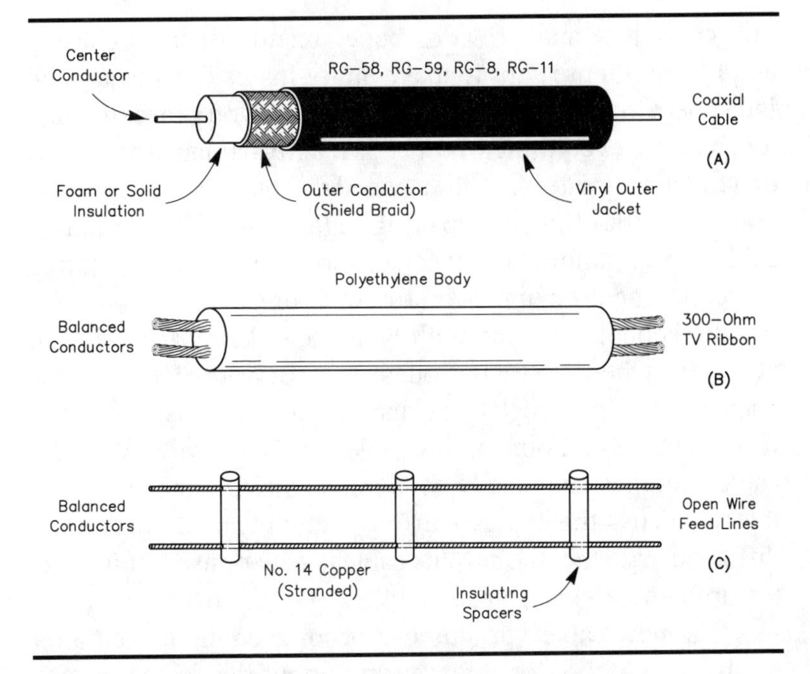

Fig 3-3—Examples that show the structure of three types of feed line. The coaxial cable at A has one conductor surrounded concentrically by the other. Feeders B and C are balanced lines with the conductors side by side and of the same diameter. The open-wire line at C has the lowest loss.

earth-moon-earth (EME) or "moonbounce," satellites or other challenging modes, you need every advantage you can get. The most successful EME operators, for example, use power amplifiers capable of the full 1500-W legal amateur power limit and with Hardline cable feeding steerable multi-antenna directional arrays that may produce an effective radiated power of more than a million watts! You can bet that they wouldn't dream of wasting half a dB to heat up a run of coax.

No matter what bands you operate on, it's generally less expensive to produce higher output power and to pick up weaker received signals with good gear *outside* the shack. This means that an efficient antenna system and top-quality feed line costs less and provides better results than purchasing amplifiers and preamps to make up for losses in antennas and feed line. Good antennas and feed line give you peace of mind, too, because everything won't be as readily damaged by storms or a rugged climate. A well-designed and built system can save you the extra effort of repairing, adjusting and maintaining cable and aluminum, so that you can spend more time operating and experimenting in the comfort of your shack.

Be economical, but don't be a cheapskate! Try to install new cable when you set up your station. Beware of old, surplus coaxial cable. Although it may have an attractive bargain price, it could be lossy. Some of this cable dates back to WW II. Old cable becomes chemically contaminated over time and when that happens, the losses can be phenomenal. It makes no difference whether the surplus cable is new or used. If it's old, it's probably not worth using.

The new cable you buy can become contaminated after it's been outdoors for a few years. Impurities in the air and ultraviolet rays take their toll on coaxial cable. The situation becomes worse when you bury coaxial line in the ground. Soil acids and alkalinity (especially in damp soil) poison the insulation in the cable and cause it to be lossy. There are special types of RG-8 cable available that are impervious to this deterioration. There are brands of impregnated line called VB-8, Impervon and others that are especially suited to in-ground use. Some types are self-healing: If a rodent chews into it, the line "heals" itself and prevents moisture and dirt from getting under the outer jacket.

Balanced Feed Lines

Coaxial cable is *unbalanced* because it has concentric inner and outer conductors. Many hams use *balanced* feed line, which has conductors side by side and of the same diameter. Standard 300-Ω TV ribbon line is an example of balanced transmission line. Another common balanced line for amateurs is known as "450-Ω ladder line." It's made commercially and has polyethylene insulation over and between the two wires.

Some hams make their own open-wire balanced feed line. The diameter of the wire and the spacing between the wires determines the line impedance. It's easy to make up your own 300-Ω, 450-Ω or 600-Ω open-wire line. The two conductors are held apart by insulators (see Fig 3-3C).

Open-wire feeders exhibit the least loss of the many feed lines we use. If we compare RG-58 cable with a 100-foot length of open-wire feed line, we find that the loss at 28 MHz through the open-wire line is only 0.17 dB, as opposed to 2 dB for the coax. This means that if your transmitter delivers 100 W to the open-wire feeder, 96 W will reach the antenna. In this case, a 4-W loss is too small to be significant.

It's not critical to have the balanced feeder's impedance precisely match the feed impedance of the antenna, but you have to take care in choosing the correct type of feed line to minimize losses. Imagine that you're using 300-Ω flat TV-reception wire as transmission line to a center-fed full-wave antenna (two ½ waves in phase). The feed impedance of this antenna will be a few thousand ohms and this causes a large mismatch at the antenna feedpoint to the 300-Ω receiving line. If you use an antenna tuner to match the overall antenna system, the SWR between the transmitter and the antenna tuner will be 1:1, so the transmitter will be happy. The losses are higher, however, in the 300-Ω TV line between the antenna tuner and the antenna, owing to the solid polyethylene insulation in which the two wires are imbedded. A 300-Ω line designed for TV reception changes its characteristics when rain or snow is present; larger, tubular 300-Ω line designed for transmitting doesn't.

An 80-meter dipole that uses open-wire 450-Ω transmitter-type feeders is an excellent multiband antenna. A balun transformer should be used between the feed line and the antenna tuner to convert from a balanced to an unbalanced condition. Many homemade and commercial tuners have a built-in balun transformer. If your tuner doesn't have a balun transformer, you should install one. Even if your low-budget tuner has a built-in balun, you may be able to get better results by bypassing the internal one and installing a better balun external to the tuner. Multiband antennas with tuned feeders are described in *W1FB's Antenna Notebook*.

Balanced feed line has its annoyances. It needs to be supported on insulated posts when long horizontal runs of line are required. If it isn't supported every 10-15 feet, it can whip in the wind and break. The wires may become crossed and short circuit the feeder. I'm referring particularly to open-wire line that has no insulation over the wires.

Vertical runs of balanced feeder can be stressed by the wind. This leads to broken feeder wires at the antenna feed-point. This potential failure can be minimized by supporting the balanced feeders every 6-10 feet along the side of your tower or mast. Use TV standoff insulators or other insulators of good quality. Stranded wire is best for balanced feed line because it's more resistant to stress breakage than is solid copper wire. I use no. 14 stranded copper wire for my home-made open-wire line.

Grounding Your Station for Safety

The most important thing to do before operating your equipment is to ensure that your gear is connected to an earth ground. This type of ground is different from the one we discussed for use with vertical antennas. Although some ground screens may be good as dc grounds for your station, others don't offer the safety against shock that you're concerned about. Electrical energy in the form of ac and dc voltage can appear on the chassis of your station gear when circuit faults develop. If the chassis and cabinets of your equipment are grounded properly, the fuses will blow when one of these faults occurs, thereby protecting you. The cabinet of each ac-operated unit you use should be connected securely to the station ground. This may be done with the shield braid from a piece of RG-58 or RG-8 coaxial cable. This ground system will help to prevent your equipment from carrying RF energy that can "bite" you when you touch the cabinets, your mike or your key when you transmit. This dc ground will help minimize malfunctioning of your radio equipment that can be caused by stray transmitter RF energy affecting the circuit performance—a common problem with solid-state gear that's often susceptible to "RF in the shack."

Good amateur practice calls for the use of one or more 8-foot ground rods driven into the earth. To reduce the possibility of electrical shock, this earth ground should be bonded back to the service entrance grounding electrode. Copper rods are best, but copper-coated steel rods may be used. I've used ¾-inch iron plumbing pipe for ground rods. It worked effectively for me. I installed a pipe cap on the top end before driving the pipe into the ground with a sledge hammer.

Once your dc ground is in place, you may connect it to the ground terminal in your ham station by routing a section of coaxial cable shield braid between the ground rods and the operating position. Keep this lead as short and direct as possible.

If your home has *all* copper plumbing, you may improve the ground system by connecting the cold-water pipes to the station ground with another run of shield braid from RG-8 cable. Iron water pipes are seldom effective as an earth ground, owing to the pipe-joint compound that seals the joints. This material, and Teflon plumbing tape, can act as an insulator to spoil the conductivity (electrical) of the overall cold-water pipe system.

Building Dipole Antennas

Now that we've discussed details about feed lines, safety and ground systems, let's talk about practical antennas you can build. We'll begin with the most common of the amateur antennas, the simple dipole. Dipoles, or doublets, are ½ wavelength long at the operating frequency. They're fed at the center and each half of the antenna is ¼ wavelength long. Dipoles have a figure-8 radiation pattern off the broad side of the antenna. This occurs only when the dipole approaches or exceeds a height of ½ wavelength above ground. Dipoles that are at ¼ wavelength or less in height radiate equally well in all directions.

Dipoles are erected in three popular ways—horizontal, vertical or sloping at a 45° angle. Vertical and sloping dipoles exhibit a virtually omnidirectional radiation pattern, as does a ¼-wave vertical antenna. A fourth and common method of erection is called an "inverted V." The center of the dipole is high above ground and the ends slope at 45° to a low height, but high enough off the ground so people can't touch the antenna and receive RF burns.

A dipole has no gain (unity) because it's used as a reference for other antennas that have gain. The term "dBd" is the gain of a given antenna, in decibels, as referenced to an ideal dipole in free space. Don't confuse this with "dBi," which refers to an isotropic antenna, a theoretically perfect spherical antenna in free space. Inflated gain figures appear in antenna literature if the manufacturer uses dBi rather than dBd to express the antenna performance characteristics and doesn't bother to mention this fact. The gain of amateur antennas is most often referenced to a dipole, which must be mounted at the same height and over the same ground as the antenna under test.

Fig 3-4 shows the physical details of a horizontal HF dipole. If the antenna were dimensioned for use at 3.7 MHz, it

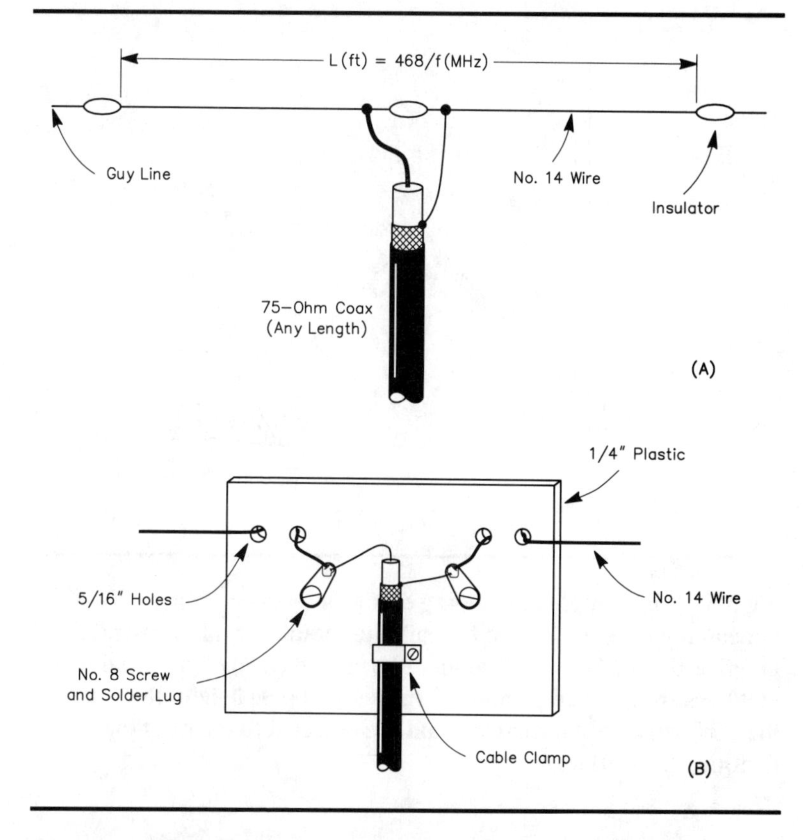

Fig 3-4—The drawing at A shows how a horizontal dipole is configured. The coaxial cable center conductor connects to half of the dipole and the shield braid is attached to the remaining half of the antenna. Drawing B illustrates the details for a homemade center insulator suitable for dipoles. A piece of oak wood may be used if it's first boiled in canning wax to weatherproof the wood. The holes through which the wires pass should be countersunk to provide tapered-hole entry. This reduces wire stress from sharp edges.

would have an overall length of 126 feet, 6 inches, in accordance with the formula in Fig 3-4A.

Try to route the feed line away from the dipole at a 90° angle. This will prevent the feeder from disturbing the

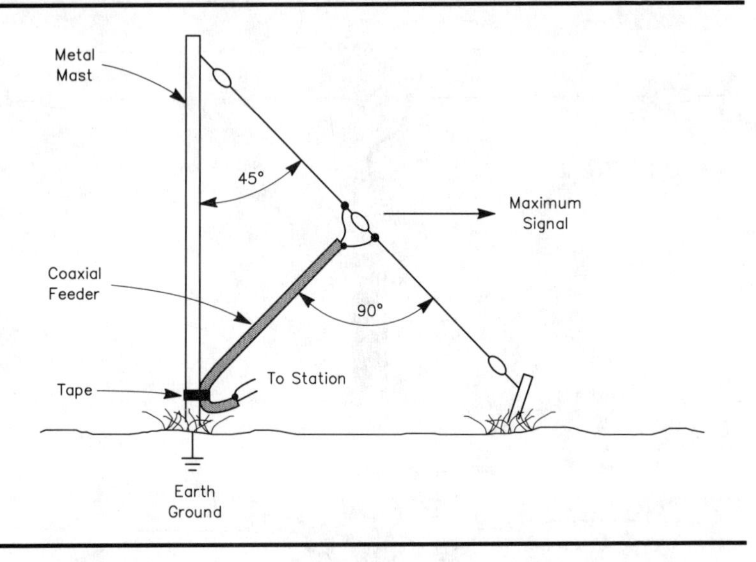

Fig 3-5—An example of a sloping dipole or "sloper." If a nonconductive mast is used instead, the radiation will be essentially nondirectional. The feed line should be routed away from the dipole at 90°, as shown. The greater the mast and dipole height, the better the performance. An earth ground is connected to the mast for lightning protection.

pattern of radiation by maintaining electrical symmetry. Many constructors will place a so-called choke balun at the antenna feedpoint to prevent stray radiation caused by currents coupled from the antenna back onto the outer shield of the coaxial cable that feeds the dipole. Such stray radiation causes the pattern to be distorted, although the effect isn't drastic if the dipole is mounted close to the ground where its pattern is essentially omnidirectional anyway. If your dipole is high above ground, orient it to favor the directions of primary interest and use a balun to preserve the pattern. For example, you may want your figure-8 radiation pattern to favor Europe and Australia; if you live in the US, you'll want to erect the antenna broadside to the NW and SW.

If your dipole is sloped toward ground from a tower or other metal support structure (see Fig 3-5), slope the dipole toward the compass point of interest. Sloping dipoles supported by metal masts of towers tend to be mildly directional in the direction of the wire slope because the tower acts as a reflector. However, you can still work stations in all directions from your site, even though your signal is strongest in the direction of the dipole slope.

The formula expressed in Fig 3-4A is an approximate one. It differs from the free-space factor (492) we use for determining height above ground. The 468 factor is necessary to compensate for what's known as antenna "end effect." This means that capacitance from the ends of the dipole to ground and nearby conductive objects changes the resonant frequency, and hence the shortening factor. If we used 492 as the factor, our dipoles would be too long to establish resonance at the desired frequency.

Even when we use 468/f (MHz) for finding the dipole length, we may need to trim some wire from the ends of the antenna. This may be the case if you erect your dipole close to ground or near trees and other detuning objects. Antenna resonance is indicated (no matter what SWR reading) by the frequency at which the SWR is the lowest. In other words, you can check operation at various frequencies in a specified ham band by observing your SWR meter. The lowest reading indicates that your antenna is purely resistive at that frequency and this identifies the resonant frequency of the antenna. If, for example, your lowest SWR reading occurs lower in frequency than the intended frequency, trim small amounts of wire from the dipole ends and keep rechecking the SWR until the lowest reading is at your chosen antenna frequency. If the lowest SWR is above the desired frequency, lengthen the dipole slightly. Don't be afraid to experiment!

I mentioned earlier the matter of dipoles being erected vertically. This provides the desired low-angle radiation for

DX work and an omnidirectional radiation pattern (vertically polarized). From a practical point of view, it's challenging to erect a vertical dipole for the amateur bands below 14 MHz. At 3.5, 7 and 10.1 MHz, it's necessary to have a rather tall supporting structure for a completely vertical dipole. Not all amateurs are able to provide a tower or mast of the required height. Furthermore, the ideal supporting device should be nonconductive (wood), which is generally impractical for mast heights above 30 feet.

Sloping dipoles are the usual choice for those of us who desire the general operating characteristics of a vertical dipole. These slopers may be used with fairly short towers or masts for 80, 40 and 30 meters. Because the sloping dipole tilts away from the mast, the metal supporting device has minor effect on the antenna performance. As I mentioned, maximum directivity occurs off the slope of the antenna and there's a slight pattern null off the back side of the tower. The radiation pattern becomes somewhat cardioid with this style of antenna.

Inverted-V HF Antennas

The drooping doublet or inverted-V dipole is one of the most popular simple antennas used by amateurs. As is the situation with sloper antennas, the inverted V needs only one support. The polarization of this antenna is mainly vertical, with a horizontal component. The radiation pattern is fairly omnidirectional. The greater the antenna height, the lower the radiation angle (best for DX work). Physical details are shown in Fig 3-6.

The recommended enclosed angle between the legs of the dipole is greater than 90°. If the enclosed angle exceeds approximately 110°, the antenna behaves like a horizontal dipole. Enclosed angles less than 90° aren't recommended because some cancellation will occur and the signal level will be reduced.

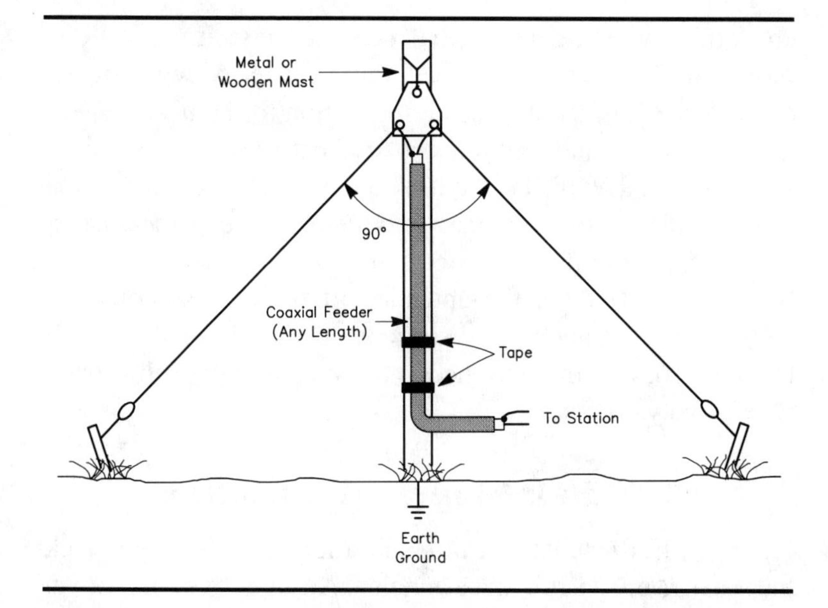

Fig 3-6—An example of an inverted-V antenna that uses coaxial feed line. RG-58 or RG-8 (50 Ω) provides a close match to this style of antenna. The coaxial feeder may be replaced with balanced, tuned feed line for multiband operation. Balanced feed requires a 4:1 balun transformer at the ham shack end of the line and an antenna tuner to obtain a 1:1 match between the feed system and the transmitter.

You'll discover L(ft) = 468/f(MHz) doesn't work well for inverted-V antennas. This is because the ends of the dipole are close to ground and to one another. Use the above formula, then add an extra foot of wire to each half of the dipole. This will allow you to shorten the dipole to provide a low SWR at the preferred operating frequency. It may be necessary to lengthen or shorten the inverted V to make it resonant at the chosen frequency. This will depend on how close the lower ends of the antenna are to ground and to nearby conductive objects. The enclosed angle will help determine the final dipole dimensions. The resonant frequency of the system isn't a vital concern if you use tuned, balanced feeders. Low SWR is important, however, if you feed your inverted V with coaxial cable. Fig 3-6 illustrates the configuration for an inverted V.

Try to keep the feed line centered between the two halves of the inverted V. It's best to have a metal mast or tower at the exact center of the antenna. This helps to keep the system balanced and nullifies the effects of a metal support and feeder being in the field of the antenna. Strive for a symmetrical installation. The antenna wire size isn't critical; use any gauge from 10-16, and the wire need not be bare copper. Insulated wire is okay, too.

The Half-Sloper HF Antenna

Fig 3-5 illustrates a full-sloper dipole. Perhaps you lack the space to erect a full-sized antenna. An attractive alternative, when space is limited, is the "half sloper." Fig 3-7 shows an example of this antenna. The radiator is a ¼-wavelength section of wire that slopes toward ground from the top of a metal mast or tower. A coaxial feed line is used (50 Ω) with this system and the feedpoint is at the top of the mast, as shown.

The shield braid of the feed line in Fig 3-7 is connected to the mast near the feedpoint and again at the bottom end of the mast, as shown. The mast is a working part of this antenna. Therefore, you need to use at least an 8-foot ground rod at the base of the mast or tower. A buried radial-wire system will improve the antenna performance, but it isn't mandatory. Any wire gauge between no. 10 and 16 is suitable for this antenna. You may use insulated or uninsulated wire.

The half sloper has characteristics similar to the full sloper in Fig 3-5. Radiation is essentially omnidirectional, with some directivity off the slope of the wire. Vertical polarization prevails. I've found this antenna effective for DX work on 160, 80 and 40 meters.

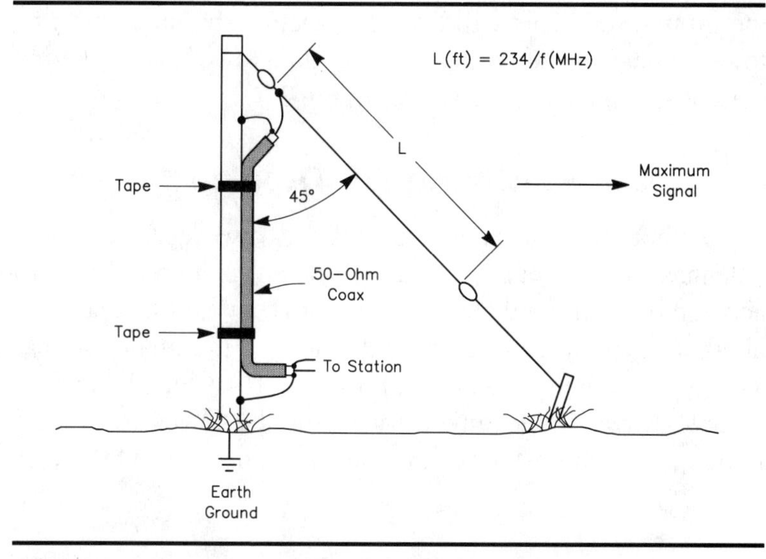

Fig 3-7—An example of a half sloper or ¼-wave sloper antenna. A metal support mast or tower is used. The shield braid of the feeder is connected to the mast near the feedpoint and at the grounded end of the mast, as shown. Maximum directivity is off the slope of the ¼-wavelength wire.

The enclosed angle of the Fig 3-7 antenna, plus the wire length, is varied experimentally to obtain the lowest SWR at the desired frequency. An SWR of 1:1 should be possible when using 50-Ω coaxial feed line. The presence of a beam antenna atop the tower will have an effect on the sloper, but you should be able to adjust and use it on towers that contain beam antennas and rotators. The performance will be affected significantly if other slopers are supported by the same tower. The additional antennas may make it impossible to obtain a low SWR. Your sloper shouldn't be erected adjacent to guy wires because they'll affect the antenna resonance.

If you don't have a metal support device for your half sloper you may use a tree or wooden mast. In this situation, you'll need to use a drop wire from the antenna feedpoint to the ground rod. This will take the place of the metal mast or tower. Consider the wire as you would a tower and make the electrical connections as shown in Fig 3-7.

Multiband HF Dipoles

Although some amateurs prefer to use separate dipole antennas for each band, you may lack sufficient real estate to accommodate more than one or two such antennas. A popular alternative to a yard filled with dipoles is a multiband dipole. This type of antenna can be built for 160-10 meters. It consists of a ½-wave dipole dimensioned for the lowest operating frequency of interest. For example, you can build an 80-meter

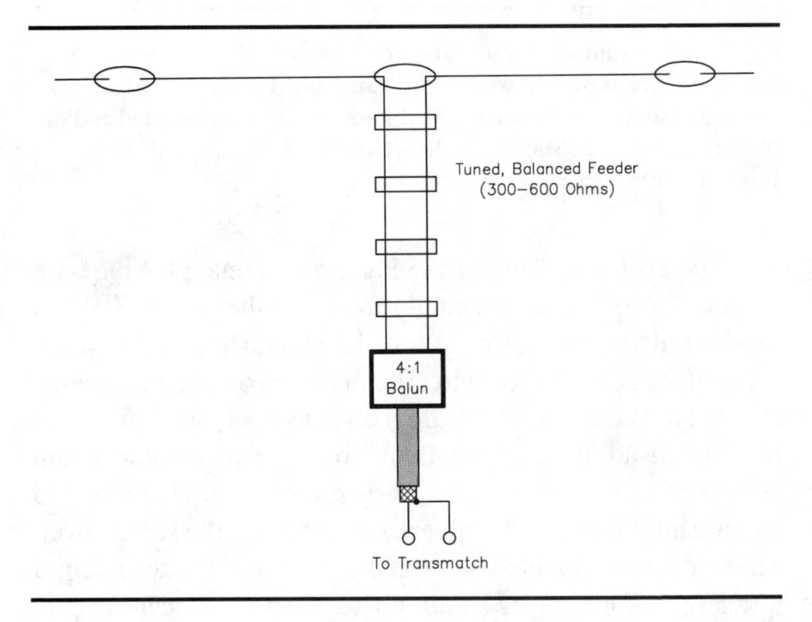

Fig 3-8—A multiband dipole with balanced feed line. The 4:1 balun transformer converts the balanced to an unbalanced line to permit using an antenna tuner designed for coaxial cable. Some commercial tuners contain a balun transformer, in which case the balanced feeders connect directly to the tuner.

dipole and substitute a tuned, balanced feed line for the coaxial feeder. Details for this multiband antenna (sometimes called a center-fed Zepp) are given in Fig 3-8. This antenna can be used effectively on 80, 40, 30, 17, 15, 12 and 10 meters if you have an antenna tuner.

You can use 300-Ω transmitting-type ribbon line, 450-Ω ladder line or home-made open-wire line for the antenna shown in Fig 3-8. Open-wire line has the lowest loss factor. Next comes 450-Ω commercial ladder line. Receiving-type TV ribbon has the greatest loss per 100 feet and is the most susceptible to changes caused by rain or snow.

The length of this multiband dipole is determined by the standard formula L(ft) = 468/f (MHz). It's not necessary to adjust the length for exact-frequency resonance when tuned feed line is employed. Your antenna tuner will provide an SWR of 1:1 when you adjust it for each operating frequency.

Another style of multiband dipole is depicted in Fig 3-9. This old technique is used successfully by many hams. It

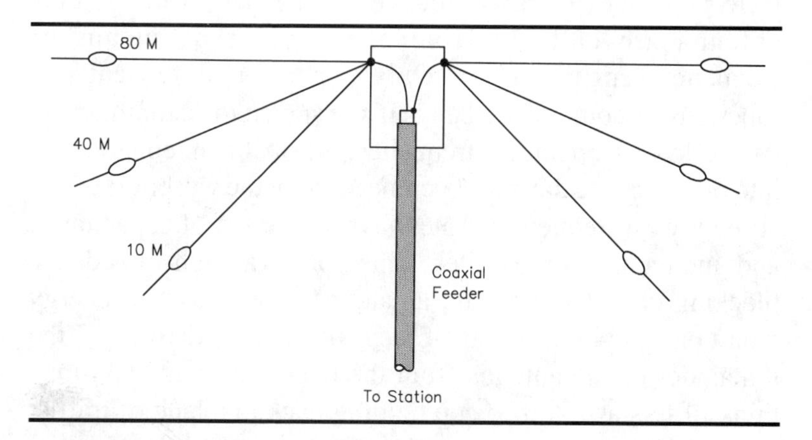

Fig 3-9—An example of a 3-band dipole that uses a single feed line. The antennas are fanned out to minimize interaction. Additional dipoles may be added or only two may be used. Norm Bradshaw, W8EEF, of St Joseph, Michigan, uses an arrangement like this for 160, 80 and 40 meters with a high degree of success.

involves using a single coaxial feeder with dipoles cut for various amateur bands. Each dipole is connected to a common feedpoint, as shown. The dipole wires are fanned apart to minimize interaction. The closer they are to one another, the greater the detuning effect for any one dipole. Therefore, the greater the distance between the wires, the easier the antenna is to adjust for resonance.

Use an SWR indicator when you adjust the dipoles for the lowest SWR. Adjust the lowest-frequency dipole first, then the dipole for the next highest frequency, and so on. This antenna has the same bandwidth limitations as a single-band dipole. At 80 and 40 meters especially, you may not be able to cover an entire band without the SWR exceeding 2:1 in some part of the band. An antenna tuner can be used to provide a 1:1 SWR for the transmitter in those parts of the affected bands.

HF Multiband Trap Dipoles

Trap antennas are popular for multiband operation. Here's how they work: Various sections of a dipole or vertical antenna are isolated from one another by means of frequency-sensitive traps. Only a portion of the antenna functions on some of the bands it's designed to accommodate. At the lowest operating frequency, all of the antenna comes into play because the traps become a part of the working system at the lowest frequency. Antenna traps consist of capacitance and inductance in parallel. This forms a high-impedance blocking circuit for certain amateur frequencies. RF energy won't pass beyond the trap of interest, thereby "divorcing" the remainder of the antenna from the overall system. You may think of this as a traffic cop holding back one lane of traffic, while allowing another lane of traffic to pass. This is a form of selective blocking.

Fig 3-10 shows how a trap dipole is arranged. Illustration A provides a look at the electrical circuit for this antenna. B is

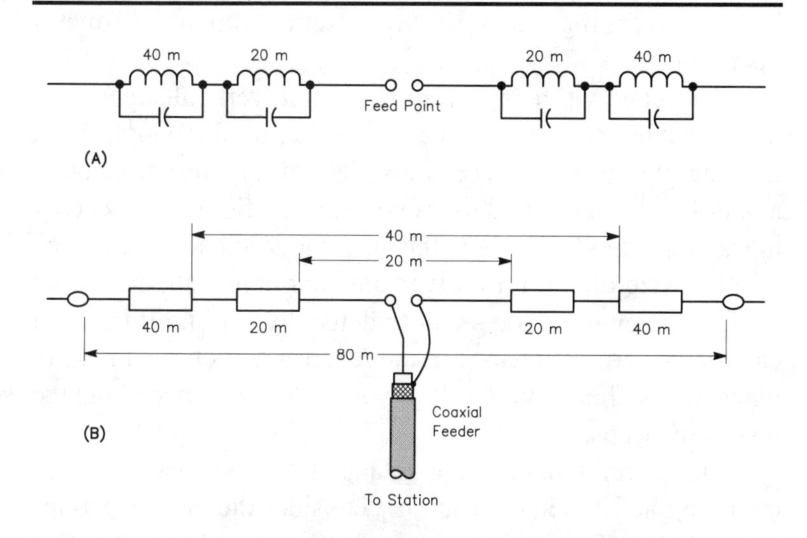

Fig 3-10—The schematic diagram at A shows coils and capacitors in parallel to form high-impedance traps that block the passage of RF energy on a selective basis. Pictorial diagram B illustrates how the traps discriminate against one frequency while passing another frequency.

a pictorial example of a three-band trap dipole. The arrows show the parts of the system that are operational versus the operating frequency.

Additional traps may be added to the antenna in Fig 3-10 to allow operation on more amateur bands. Fig 3-10 shows that the traps that aren't blocking the flow of energy become a working portion of the antenna at the frequencies below that of the specified trap. If you're operating in the 40-meter band, the 20-meter traps allow the 40-meter energy to pass until it reaches the 40-meter trap. In this example the 20-meter trap acts as a loading coil on 40 meters. Therefore, the overall length of the 40-meter portion of the antenna is shorter than it would be if it were a single dipole for 40 meters. All of the traps become working parts of the antenna during 80-meter operation.

The bandwidth of a trap dipole or vertical antenna is somewhat less than we can realize from a full-size, single-band antenna. We're able to accept this trade-off in return for having a multiband antenna that uses one feed line. Some losses occur in the traps and this lowers the antenna efficiency below that of a full-size dipole for a given amateur band. The reduction in signal strength is too small to detect by ear and it may not cause a discernible change in the receiver S-meter reading. In other words, there's no need for you to be concerned about the loss in the traps.

Please refer once more to Fig 3-10B. Suppose you're operating at 20 meters. You may consider the 20-meter traps as insulators. The additional wire beyond the 20-meter traps, plus the 40-meter traps, is electrically "divorced" from the part of the antenna between the feedpoint and the 20-meter traps. Each trap presents a high impedance many thousands of ohms at the operating frequency. This resistance is known as reactance in ac and RF circuits. This treatment of how traps operate may seem repetitive, but trap-antenna operation is baffling to many amateurs. I want to be certain that you're aware of what's happening within this type of antenna before we move along to other antennas.

The major shortcoming of a multiband trap dipole is its weight. The greater the number of traps the more it weighs. This places stress on the antenna wire and it can cause the dipole to sag. The inverted-V configuration is less stressful to the system than the horizontal dipole format because it's supported at the center. The exception for the latter condition is when the horizontal trap dipole is supported at its ends and at the center.

Antenna traps need to be protected from rain, snow and dirt. If you make your own traps, it's prudent to enclose them in plastic tubing. Rain, snow and ice detunes the traps and leads to impaired antenna performance. An accumulation of dirt can degrade the trap performance. Information about constructing and tuning your own trap dipole is provided in the *ARRL Antenna Book* and in *W1FB's Antenna Notebook.*

Vertical Antennas: HF and VHF

If you live where antenna space is at a premium or if you plan to operate mainly VHF/UHF FM, it will be advantageous for your antennas to go "up" rather than "out." Vertical antennas are often the choice of urban or apartment dwellers for the reason just mentioned. VHF/UHF FM signals are almost always vertically polarized, so a horizontal antenna will knock your signals down 3 dB or more when working FM simplex or into a repeater. At HF, a ground- or roof-mounted vertical can outperform a dipole that's not high above ground. This is especially true if you're a DX-oriented person. Vertical antennas have a low angle of radiation, which is ideal for HF DX work, and they're omnidirectional, which eliminates the need for an antenna rotator.

A vertical antenna may consist of a piece of wire or it may be made from tubing, pipe or tower sections. The greater the antenna cross-sectional area, the greater the effective bandwidth. Hence, a tower-type vertical has greater bandwidth than a vertical made from a piece of no. 12 wire. This is because the Q (quality factor) becomes higher as the antenna diameter decreases. The lower the Q, the greater the bandwidth.

The disadvantages associated with ¼-wavelength HF verticals are the need for a ground screen or counterpoise and the ease with which they respond to man-made noise on receive. Because most man-made noise (QRN) is vertically polarized, this antenna picks up the noise because it's also vertically polarized. This may not cause a problem if you live in a relatively "quiet" neighborhood. We discussed various types of ground screens earlier in this chapter, so we won't treat

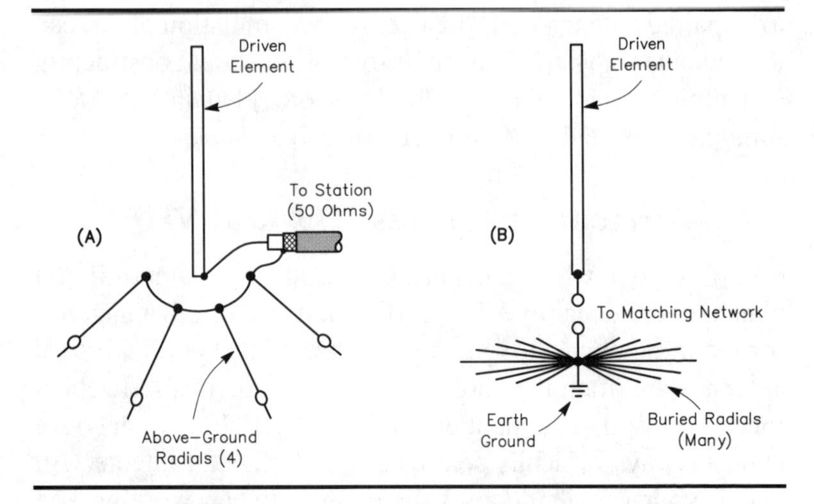

Fig 3-11—Examples of 1⁄4-wavelength vertical antennas. The antenna at A has above-ground radials approximately 5% longer than the driven element. The length of the driven elements for both antennas is obtained from L(ft) = 234/f(MHz). The radial length at A can be calculated from a similar formula: L(ft) = 246/f(MHz). If the radials are spaced 90° apart and drooped at 45° (antenna A) the feedpoint will closely match 50-Ω coaxial cable. If the radials are parallel to ground the impedance will drop to approximately 30 Ω. Antenna B is a ground-mounted vertical. The radials are buried in the earth or laid on the ground. They can be any length, but should be (ideally) 1⁄4 wavelength long for a full-size driven element. The feed-point impedance for antenna B is low—a few ohms to approximately 30 Ω, depending on the length of the driven element radiator.

that topic again, except to say that an above-ground vertical antenna can be made to work quite well with only four radial wires. These wires may be used as guy lines for the system if the lower ends of the radials are equipped with insulators. Fig 3-11 illustrates two common types of vertical antennas. They are shown as single-banders.

VHF and UHF verticals are relatively short, however, and rarely require guy lines. Three or four radials (wires or metal rods) are usually sufficient to provide the proper ground plane. The antenna's impedance is partly determined by the method of mounting the radials. A ⅝-wavelength whip, for example, presents close to a 50-Ω impedance match with three or four radials bent downward at about a 45° angle.

The antenna in Fig 3-11A is easy to use with 50-Ω coaxial cable if the radials are dropped at 45°. This isn't the case with the antenna of Fig 3-11B. It has a lower feed impedance and requires a matching network to ensure a low SWR. Suitable matching networks for verticals are described in the *ARRL Antenna Book.*

You may use traps to obtain multiband operation with your vertical. Numerous commercial antennas of this type are available to amateurs. For HF operation, the installation for an above-ground trap vertical becomes somewhat complicated because the system requires at least two radial wires for each band of operation. Antenna adjustment for the lowest SWR on each amateur band can be a tedious undertaking when several traps are used. Nonetheless, these short multiband antennas are used by many hams who lack the space for larger antennas. The higher above ground you erect your vertical antenna, the better the performance. Most multiband ground-mounted verticals don't offer outstanding performance because nearby conductive objects affect the radiation pattern and absorb RF energy. I don't recommend a ground-mounted single- or multiband vertical for any frequency above 7 MHz.

Big Loop Antennas for HF

If you have a substantial amount of real estate, consider erecting a full-wave loop antenna. It can be any convenient shape, although the greatest gain occurs when the loop is perfectly circular, a configuration that's not easy to make. Next is the common square loop, followed by the delta (triangular) loop. A rectangular loop provides the least gain of the group.

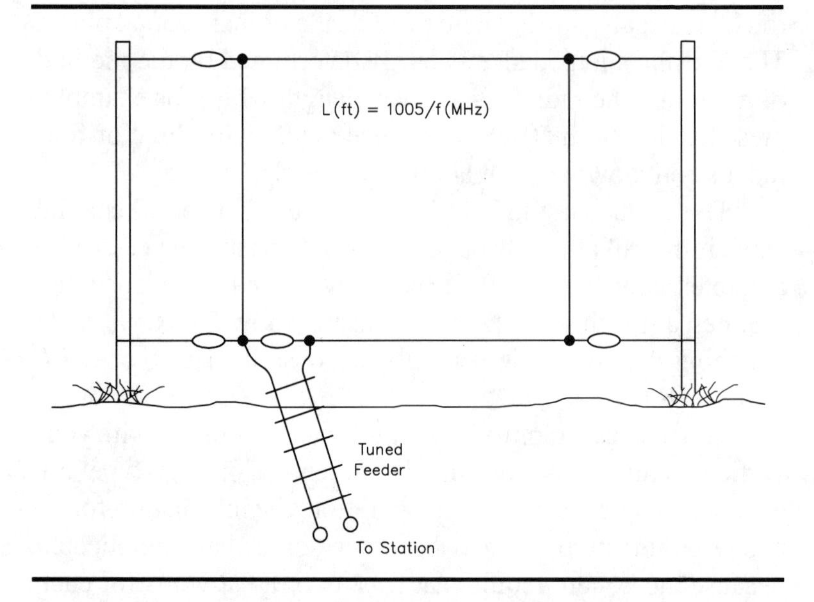

Fig 3-12—An example of a full-wave loop for multiband operation. The feed line can be 300-Ω transmitting ribbon, 450-Ω ladder line or homemade open-wire line. An antenna tuner and 4:1 balun transformer is used at the transmitter to tune the system for the desired band of operation.

The more rectangular it becomes, the lower the gain. The approximate overall dimension for a full-wave loop is derived from Length (ft) = 1005/f (MHz). Thus, if you want to build a loop for 40 meters (7.1 MHz), use 141 feet, 6½ inches of wire. If the loop is square, it will have 35 feet, 4½ inches of wire per side.

A large loop is a low-Q radiator. Because of this, it will have greater bandwidth than a dipole cut for the same frequency and it's not affected as significantly by nearby conductive objects as is a dipole or vertical. Perhaps the greatest benefit derived from the use of a loop antenna is the low-noise reception that results. This is because a closed loop isn't as receptive to man-made QRN as are a number of other amateur antennas. This "quiet" antenna is excellent for weak-signal reception, such as may be required when working DX.

The full-wave loop can be used for several bands of operation by using tuned, balanced feeders. Fig 3-12 shows how this may be done with a square loop. The feed method applies equally well to a loop of any shape.

The loop in Fig 3-12 may be used from its fundamental frequency through 10 meters. Therefore, if you design it for 80 meters you can use it from 3.5-29 MHz. If you prefer to build this antenna for single-band operation, simply replace the tuned feeders with coaxial cable. To obtain an impedance match when using coaxial line (the loop impedance is approximately 115 Ω) it is necessary to install a ¼-wavelength coaxial matching section at the feedpoint. Use 75-Ω coaxial cable for this device. The length is obtained from 246/f(MHz) times the velocity factor of the coaxial line you're using. The velocity factor for most nonfoam dielectric 50 or 75-Ω cable is 0.66. If we want to make a matching section for a 7.1-MHz loop. (L (ft) = 246/7.1 MHz. This results in a length of 34 feet, 7¾ inches. Now multiply this number by 0.66. The resultant cable length is 24 feet, 6½ inches. This matching section is used between the loop feedpoint and the 50-Ω coaxial cable to the ham shack. The 50-Ω feeder may be any convenient length.

Loop antennas may be fed at various points along the wire. The example in Fig 3-12 is fed at a lower corner to provide low-angle radiation and vertical polarization. This is a convenient spot to connect the feed line, but slightly better performance will result if we feed the loop at the center of one of the vertical sides. You can obtain horizontal polarization by feeding the antenna at the center of the top (preferred) or in the center of the bottom wire. Your loop may be delta shaped if you have only one support structure. The tip of the triangle is attached at the top of the mast and the flat side of the triangle (opposite the tip) is made parallel to ground. The antenna can be fed at the tip, in one corner or in the middle of the lower side. If the mast isn't high enough to support the antenna vertically, tilt the bottom of the loop away from the mast.

The higher the loop is above ground, the better it will perform. Although I've used several large loops that had the lower sides only 5-6 feet above ground, they didn't offer the DX performance I enjoyed with loops that had the bottom wires 25-40 feet above ground.

What about Horizontal Loops for HF?

The question about loops lying completely parallel to ground is asked frequently. I'm using a 160-meter square loop parallel to earth and only 50 feet high. This type of antenna is excellent for close-in communication. It beams the energy upward toward the ionosphere, rather than at a low angle to the horizon. It's a fine antenna for distances out to 500 or 600 miles on 1.8 MHz. The antenna is unsuitable for DXing. I used tuned 450-Ω feeders to allow multiband operation. The antenna tuner contains a 4:1 balun transformer that changes the balanced feed to unbalanced coaxial line. The antenna has a lower angle of radiation as the operating frequency is increased because the effective height increases with frequency. At 80 meters, the system is effective out to 1000 miles or greater, and at 40 meters it's a fine DX antenna. Likewise on 20, 17, 15, 12 and 10 meters.

If you install your loop vertically (perpendicular to ground), it will radiate maximum energy off the broad side of the loop at the lowest operating frequency. Maximum radiation is in the plane (off the ends) of the loop at harmonics of the fundamental frequency. The horizontal loop, on the other hand, radiates equally well in all directions.

Directional Beam Antennas

It's helpful to have an antenna that provides gain in a desired direction. The use of a rotator will enable you to beam your signal toward Europe, desired grid squares, distant repeaters, packet stations and other parts of the world. The response off the sides and the back of a properly designed beam antenna is far below the response off the front side of the antenna and it's this reduction of response in certain directions and concentration in others that produces gain. This helps to reduce interference from directions that aren't of interest. The antenna gain (referenced to a dipole) increases with the number of elements. Yagi and cubical-quad beam antennas are the most popular directional beam antennas for amateur use. Cubical quads consist of two or more full-wave loops. The director resonance is higher than that of the driven element and the reflector loop is resonant below the driven-element frequency. Yagi antennas have elements made from metal tubing or rods. Each element is approximately ½ wavelength overall. The directors are shorter than the driven element and the reflector is slightly longer. A shortening and lengthening factor of roughly 5% is used for the directors and reflector, respectively. Fig 3-13 shows how a 3-element Yagi is configured. The antenna gain is 7-8 dB over that obtained with a dipole. The gain is dependent upon the element spacing and the care used in matching the feed line to the driven element.

A great many directors are used in practical VHF and UHF Yagis, but the practice is too cumbersome for HF-band work, where the dimensions are much larger. It's possible to obtain forward gain of 15 dB or greater with a single VHF or UHF Yagi with many elements.

Yagi and quad beam antennas have a front-to-back characteristic. This defines the forward gain compared to the response off the back side (reflector) of the antenna. The front-to-back ratio depends on the spacing between the

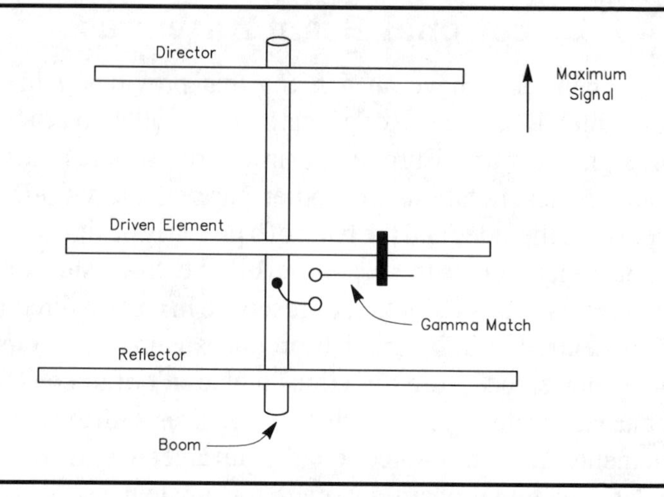

Fig 3-13—Layout for a simple 3-element Yagi beam antenna. The director is slightly shorter than the driven element and the reflector is slightly longer than the driven element (see text). A gamma matching arm is shown, but other types of matching devices are used. This device matches the feed line to the feedpoint of the Yagi.

elements and on element tuning. The element spacing has an effect on the forward gain and the feed impedance. You can learn more about these matters by reading the *ARRL Antenna Book.*

HF Multiband Trap Yagis

Perhaps the most popular all-around HF beam antenna used by hams is the "tribander." This is a Yagi constructed with three or more elements like the one in Fig 3-13, but it has traps in the elements to allow use on more than one amateur band. The principle is the same as is shown in Fig 3-10. The traps act as loading coils on some bands and this results in a Yagi that's shorter than it would be for the lowest band of operation. The antenna weight, however, increases in proportion to the size and number of traps. A typical tribander is set up for operation on 10, 15 and 20 meters. Additional traps may be used to allow operation also on 12 and 17 meters. As is the case with trap dipoles, a single coaxial feed line is used to feed a trap Yagi antenna.

You'll note in the instructions for commercial HF-band Yagis that you should use a decoupling choke (or a 1:1 balun transformer) at the feedpoint of the beam antenna. The decoupling coil is made from RG-8 coaxial cable (several turns, 6-inch OD). This coil acts as an RF choke to prevent currents from the antenna from being induced onto the outer shield of the feed line. If the feed line is allowed to radiate these stray currents it can spoil the radiation pattern of the Yagi. Large ferrite sleeves (4-6 units, typically) may be slipped over the coaxial feeder near the Yagi feedpoint to serve as a decoupling circuit in lieu of a coaxial decoupling choke. The sleeves are taped securely in place on the coaxial feeder.

Do You Need a Rotary Beam Antenna?

Beam antennas require a rugged mast or tower for support. A husky antenna rotator is required—one designed to turn big, heavy antennas. Small TV-type rotators aren't suitable for use with HF-band beam antennas, but are satisfactory for turning modest size VHF and UHF arrays. The cost of your antenna system increases when you install a rotary beam antenna. Is this a prudent investment?

Your need for a tower, beam antenna, rotator and related components depends on your operating objectives. If you enjoy casual HF communications worldwide, dipoles, loops or vertical antennas will generally suffice. If, one the other hand, you're dedicated to DXing, contesting, or reaching far-off VHF/UHF repeaters or packet stations, you'll enjoy the advantages associated with a gain type of directional antenna. If you're a newcomer to Amateur Radio, avoid investing in an elaborate antenna system until you identify your operating objectives.

Bringing Feed Lines and Control Lines into Your Shack

Most of us fret about the mechanical problems associated with the routing of feed lines, earth grounds and rotator cables that must come from out of doors and into the radio room. Some amateurs bore holes in the foundation, just above ground level, then route the wires into the cellar or basement. This is fine if you have a "studio B" shack (basement setup), but if the radio station is on the first or second floor, you'll have to route the wires up through the wall. A less attractive method calls for boring a hole or holes in the floor near the molding. I prefer a less severe approach to the problem. Examples of methods I've used are shown in Fig 3-14. Two options cause no damage to the dwelling, such as in illustrations A and B. Fig 3-14C requires cutting a square hole in the inner and outer walls of the house.

The window in Fig 3-14A needs to be secured to prevent intruders from entering the house easily. Fig 3-14B shows how an insert panel may be substituted for a window pane, then puttied for weather sealing. The inner and outer metal plates in Fig 3-14C should be sealed against the weather to prevent drafts and moisture from entering the house. The bolts for the two feed-through bushings at C serve as the conductors. Coaxial cable is used for the SO-239 inner lines. Ground the shield braid of each coaxial line to the inner and outer metal plates.

Another technique you may wish to consider involves using PVC plumbing pipe (1½ or 2-inch size) as a through-wall conduit. A PVC 90° elbow is used at the outside of the house (pointed down) to prevent moisture from entering the conduit. After the cables are routed through the conduit, it may

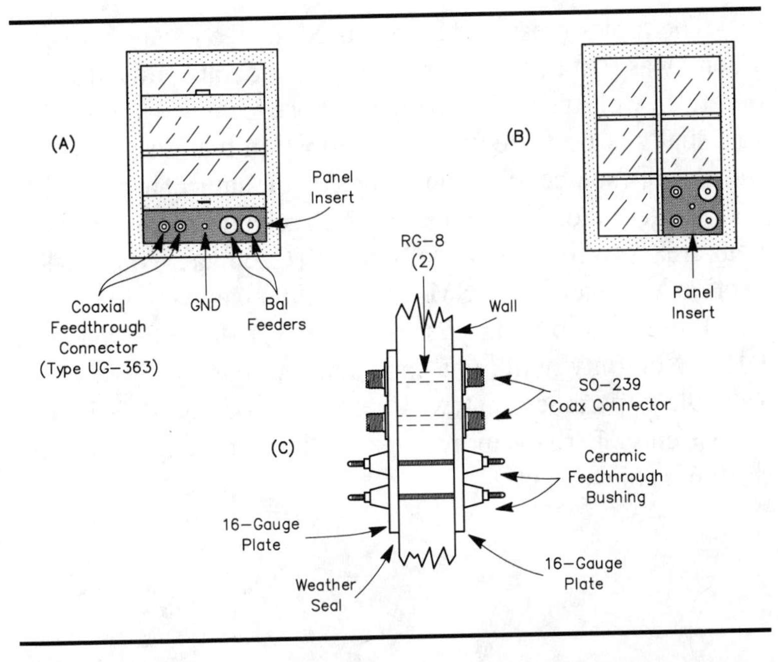

Fig 3-14—Examples of ways to bring cable and wires into the shack from outside the house. Method A uses a jack panel held in place by a raised window. Illustration B shows how to replace a glass window pane with a jack panel. The panels may be made from Masonite, Plexiglas or metal. The technique at C requires cutting a square hole in the inner and outer wall of the house. Metal plates are screwed to the walls to serve as jack panels.

be sealed with caulking to keep dirt, insects and rodents from entering the house.

Summary

I've mentioned the most basic of antenna considerations in this chapter. Expand your knowledge by reading *W1FB's Antenna Notebook*, available from radio dealers and ARRL HQ. This book is written in plain language and has many antenna designs that you can duplicate easily.

The topics presented here caused me the most confusion when I was first licensed. I received considerable misinformation from well-meaning amateurs whose technical knowledge was shaky at best. Beware of information that doesn't come from a solid source. If in doubt, look up the subject in an ARRL book or get in touch with the ARRL Technical Coordinator in your area. (To find the name of your TC, contact your ARRL Section Manager (SM). SMs are listed on page 8 of *QST*.)

I urge you to build your own antennas for HF, VHF and UHF. Not only will you save money, but you'll acquire valuable experience that will help you upgrade to a higher license class. Perhaps more importantly, you'll experience the thrill of achievement!

Glossary

Antenna—A device made from electrical conductors that radiates RF energy and absorbs incoming RF energy.

Balanced feeder—A feed line that has two parallel conductors of equivalent cross-sectional area. The spacing between these conductors is maintained at a constant value.

Bandwidth (antenna)—The effective bandwidth between arbitrary frequency limits that results in a given SWR. Generally, the frequency range between the 2:1 SWR points is considered the antenna bandwidth by amateurs.

Beam antenna—An antenna that provides gain over a dipole (dBd in free space) and concentrates the signal energy in a specific direction. A beam of energy is formed and directed toward a desired part of the country or world.

Conductor—A low-resistance electrical conductor. Wire, metal tubing and metal rod material is generally used to form antenna conductors.

Counterpoise—An electrical conductor supported above ground to serve as an artificial earth ground for use with a ¼-wavelength antenna. The counterpoise wire is usually ¼ wavelength or greater overall.

Cubical quad—A type of beam antenna that is composed of full-wave loops of wire for the reflector, driven element and director.

Dipole—Also known as a "doublet." An antenna that has two electrical conductors, each ¼ wavelength long (making a ½-wave dipole). The feed line is attached at the center of the dipole.

Directional antenna—An antenna that can be made to concentrate the radiated energy in one or more chosen directions, such as is the case for a beam antenna.

Driven element—The conductor in an antenna or antenna array to which the RF energy is supplied by means of a feed line.

Feed impedance—The characteristic antenna resistance at the point where the feed line is attached. The ac resistance at resonance where there's no inductive or capacitive reactance present.

Feed line—Also called feeder or transmission line. Coaxial or balanced cable used to conduct RF energy from a transmitter to an antenna or from an antenna to a receiver.

Feed line loss—The signal loss in decibels (dB) for a specified length of feed line. The loss is generally expressed for 100 feet of line and varies with the operating frequency.

Full sloper—A dipole antenna erected neither vertically nor horizontally. A sloper usually has a pitch angle about 45° from vertical.

Ground (artificial)—A system of conductors used in place of a true earth ground. Sometimes called a "ground screen," radial wires or counterpoise wires constitute an artificial ground.

Ground (true)—The area in the earth where current is able to flow.

Ground-mounted —An antenna (usually a vertical) installed at ground level, rather than on a mast, tower or other high supporting device.

Half sloper—Similar to a full sloper. The radiator is a ¼-wavelength wire or conductor. The metal support structure (such as a tower) becomes a part of the antenna because one conductor of the feed line is attached to the support device.

Horizontal antenna—A dipole, end-fed wire or beam antenna that has its element or elements parallel to ground to produce horizontal polarization.

Image antenna—The phantom portion of an antenna in the earth below the radiator. See Fig 3-1.

Inverted V—A dipole antenna supported at the center, but with the outer ends near ground. The dipole wires slope at about 45° to form an enclosed angle of 90°. Also called a "drooping dipole" or "drooping doublet."

Isotropic antenna—A theoretically perfect spherical antenna in free space with nothing connected or coupled to it. Used as an ideal reference for antenna performance. An isotropic antenna would, if realized, radiate energy equally in all directions.

Loop antenna—A conductor that's circular, square or rectangular. The loop is closed, except at the feedpoint.

Multiband antenna—An antenna capable of operating on several amateur bands with a single feed line.

Q (antenna)—Antenna figure of merit, or "Quality" factor. The larger the antenna conductor, the lower the Q and the greater the bandwidth.

Radials—A system of conductors deployed below a ¼-wavelength radiator to form a ground screen. One conductor of the feed line is attached to the center of the radial system.

Radiator—The conductor in an antenna system that radiates RF energy. Also called "driven element."

Trap—A resonant device consisting of paralleled capacitance and inductance. Used in multiband antennas to stop RF current flow at selected frequencies.

Yagi (also Yagi-Uda array)—Named after the Japanese professors Yagi and Uda, who first described this type of antenna in the 1920s. A directional beam antenna with a reflector, driven element and one or more directors.

Chapter 4

Station Layout and Safety

Amateur Radio stations should be comfortable, illuminated properly and above all else, *safe*! You may spend many hours a week in your ham shack and you should install your radio gear in an orderly fashion. For instance, there's less fatigue and inconvenience if equipment controls are within easy reach from your chair and there's adequate lighting to ensure that you can read the dials and panels clearly. Logkeeping and notetaking are easier with good illumination, too.

Ham stations must always be safe from electrical hazards, such as chassis and cabinets that have an ac or dc potential, and lightning strikes on towers, masts and antennas.

Radio Room Location

You may wonder which area of your home is best for setting up a radio room. It depends on available space. Attics are used as radio rooms by some amateurs. This part of the home provides good isolation from other household activities, but there are drawbacks. Attics are often hot in the summer and cold in the winter. Another minus factor is that it's a long way to earth ground from an attic, so it's difficult to obtain an effective RF ground for the station. In this case, you may find that the equipment cabinets, mikes and keys are "hot" with RF energy when you operate. A commercial "artificial ground" tuner can help in these situations because it lets you tune the ground lead for each operating frequency. This cancels the reactance in the ground wire to eliminate unwanted RF energy in the shack. These artificial-ground units don't take the place of a true earth ground, however; they simply help to make a poor earth ground more effective.

What about basement radio shacks? A cellar or basement that's heated in the cold months (and isn't damp) is fine for a station location (sometimes called Studio B, with Studio A representing an attic shack). During the summer, when basements tend to become damp, a dehumidifier can help keep the atmosphere dry. A damp basement is a poor place for electronics equipment. The moisture can cause rust and it may cause high-voltage circuits to arc. Dampness can cause mildew to form on circuit boards and circuit wiring. A room that's too dry, however, can increase the buildup of static electricity, and could make it more likely for you to accidentally "zap" your equipment when you touch it. You'll have to judge your basement as a radio room. An advantage of the basement shack is that the earth-ground circuit is usually good because of the short leads needed between the equipment and the ground rods and/or copper cold-water pipes.

A first-floor den or bedroom is a fine location for a ham shack. Second- or third-floor installations can cause problems with stray RF energy, much like an attic shack. Generally, a first-floor room is close enough to the earth ground to prevent stray RF problems. The room should have a door to isolate your activities from those elsewhere in the house. For example, loud sounds coming from your speakers can be disturbing to those who want to read, view TV or carry on a conversation. With the shack door closed, you won't be disturbed by sounds that originate outside your shack and these unwanted sounds won't be transmitted with your voice.

Make sure fans, air conditioners, dehumidifiers and other appliances don't make noises that will be picked up by your microphone. Use thick curtains or drapes, and padded furniture to reduce room noise and any "reverberation" effects that can be caused by your voice or other sounds reflecting off hard, flat surfaces. Also, make sure you have adequate ventilation. A poorly designed or malfunctioning space heater can suffocate you without warning, and you'll feel better and stay alert longer during extended operating sessions if you have plenty of fresh air. Ventilation is of greater importance if you smoke; cigar, cigarette or pipe smoking can deposit damaging chemical residue on circuit boards, components and external surfaces of electronic equipment.

Many ham stations of yesteryear were set apart from the main dwelling, in separate little buildings (hence the name "ham shack"). This is something you may want to consider when choosing a location. I've seen two California ham shacks in wooden garden sheds in the backyard. These buildings come as kits from lumberyards and it's easy to insulate them if you live in a region where winter temperatures are low. A small gas or electric heater may be sufficient to provide comfort during cold months. I've seen ham stations set up in small mobile homes and camp trailers.

Keep Your Antennas Near the Shack

Before selecting the radio room site, try to decide where the antenna tower or mast will be placed. It's best to have your shack as near to the tower or mast as practical to keep feed lines short and reduce signal losses. Also, feed lines won't have to be routed around or over the house to reach your equipment. It's helpful to choose a site close to metal cold-water pipes, which can serve nicely as a station ground. If your home has PVC pipe for plumbing, it can't serve as your ground because plastic is nonconductive.

Ham Shack Lighting

Fluorescent lamps can cause static in a ham shack. This form of manmade interference shows up as a loud buzz in the receiver output. It's possible to suppress noise in fluorescent lights by installing 0.1-µF bypass capacitors at key points in the lamp housing, if you're so inclined. Older lamps that require starters can sometimes be "tamed" by installing new starters or bulbs. Many modern fluorescent "instant-start" lamps don't cause QRN when they're working properly. You can track down a noisy fluorescent light in your home by turning off lamps one at a time until the noise vanishes from your receiver output.

Try to illuminate your radio room with incandescent lamps. These don't cause noise. I prefer a ceiling light with a 75- or 100-W bulb. This serves as my primary light source for equipment installation or repair. I use a swingaway desk lamp (clamped to the right-hand edge of my desk) with a 75-W bulb as the light source for my operating. The lamp is adjusted so that the main beam falls at the center of the desk in front of me. The hood over the bulb directs the light downward to prevent light from diminishing the digital readout displays on my equipment. You may prefer to use a small, high-intensity desk lamp instead of the large type I use. Choose a light source that suits your taste and budget.

Beware of light dimmers. Most of these devices generate wideband noise or "hash." This kind of QRN can wipe out ham band reception. The lower your operating frequency in MHz, the worse the problem. You may have neighbors who have noisy light dimmers in their homes. Don't rule out this possibility if you can't find a noise source in your own home.

Your Operating Desk or Table

Economy may be necessary in your first station. A simple, inexpensive operating table may be your preference during your first few months as a ham. A used standard office desk is worth considering. I purchased a nice wooden office desk from a used-equipment dealer for $25. Used metal desks were available in the same warehouse for as little as $50. Some had dents and blemishes, but they could have been easily cleaned and painted.

Perhaps you have an old smooth-surface door in your storage area. Most modern doors of this variety are hollow, but have a nice veneer finish. They're fine for use as desktops and they're nice and long for your big ham station. You can attach screw-on iron legs to the door or it can be supported by two small chests of drawers or filing cabinets. Look for these at used office-supply stores. Fig 4-1 shows how you can use two of these cabinets to support a desk made from a door.

Lumberyards sell Formica-covered sink countertops at reasonable prices. These serve as fine desktops. I use one of these, supported by two short file cabinets, as my desk for office work and to support my computer.

You can build an equipment shelf to place on top of your homemade desk. White pine lumber or ¾-inch plywood is suitable, and Masonite can be tacked to the top surface to improve the appearance. If you're experienced at cutting and

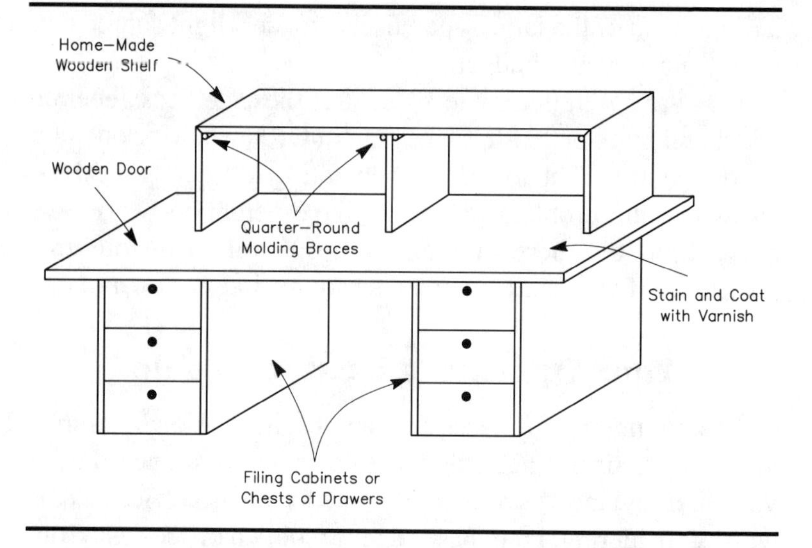

Fig 4-1—An example of how to use a wooden door and two file cabinets to form a large, inexpensive desk. The desktop shelf can be dimensioned to suit your needs.

installing Formica sheeting, consider this as a shelf or desktop. Adding a shelf to your desk permits you to place a substantial number of items above the desktop, allowing more arm room on the desk. Most of my station accessories (smaller units) are on the shelf above my desk.

Increasing the Surface Area of Your Office Desk

If you use an office desk as your operating table, you can increase the surface area easily and inexpensively. Place a sheet of ¾-inch plywood or particleboard over the existing desktop. This extends the right and left sides of the desktop. My desk has this feature; I gained two feet of extra space on the left side of my desk and a foot of new space on the right side. I screwed strips of ¾×1½-inch wood to the bottom side of the new

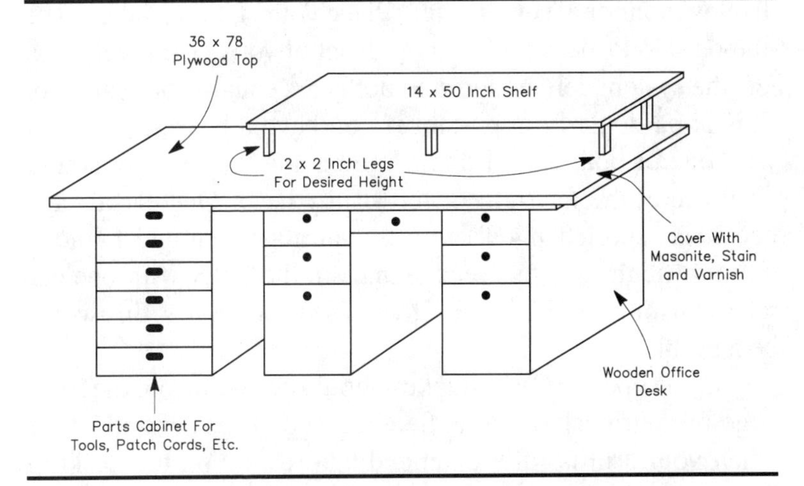

Fig 4-2—The layout of the W1FB operating position. The shelf is 10 inches high and uses manufactured wooden spindles as supports. Posts made from 2×2-inch lumber are suitable, as shown. A wooden vanity-type cabinet supports the left end of the plywood desktop.

desktop. These pieces form a border that encompasses the edges of the main desktop and prevents the new desktop from moving around. I covered the plywood desktop addition with Masonite, then added trim molding around the edges of the new top. Fig 4-2 is a sketch of my layout with the old wooden office desk.

Arranging Your Equipment

Your most important layout consideration is convenience. Place the most frequently used equipment in front of you on the desk. Your transceiver will get most of your attention if you operate phone or CW. This involves tuning up, reading the frequency display and other such considerations. If you operate a mode that uses a computer, such as packet, RTTY, AMTOR, PacTOR, G-TOR or CLOVER, you'll want your keyboard and display monitor at center stage. Place your station's inboard or outboard loudspeaker directly in front of you at ear level. This lets the system deliver audio to both ears equally, or nearly so. This is particularly important if you have a hearing problem with one or both ears. I don't like having the receiver sound originate at the far right or left of my head. One ear always becomes "shortchanged" with this arrangement and I find it more difficult to copy weak signals in the QRN with one ear receiving most of the sound. If you use headphones, this won't be a problem.

If you operate CW, your key should be away from the front edge of your desk or table. Sending code is easier and better when your arm is fully extended and resting on the desktop from your elbow to your wrist. This reduces tension on your arm muscles and contributes to smoother sending. It also reduces arm fatigue, which is a major benefit when working contests. With all the consideration given to ergonomics in workplaces today, perhaps you'll find that the comfort factor is also a health factor. It never hurts to play it safe!

If you use a keyboard for CW, it should be at a comfortable height; if it's too high or low for your chair, you'll be uncomfortable and arm fatigue may result. A computer monitor screen should be set up high enough that you can look directly forward at it when sitting fully upright. Your keyboard should be lower than standard desktop height.

I use a linear amplifier when poor band conditions require a signal boost. I keep this unit on the desktop at the far left. It isn't adjusted as often as my transceiver or antenna tuner, so I don't need it directly in front of me.

I use the desktop shelf to accommodate my antenna rotator control, a 2-meter transceiver, an RF power meter and an antenna tuner. The RF power meter/SWR indicator and antenna tuner are in front of me because they're called upon each time I operate. You can see how my equipment is arranged by referring to Fig 4-3.

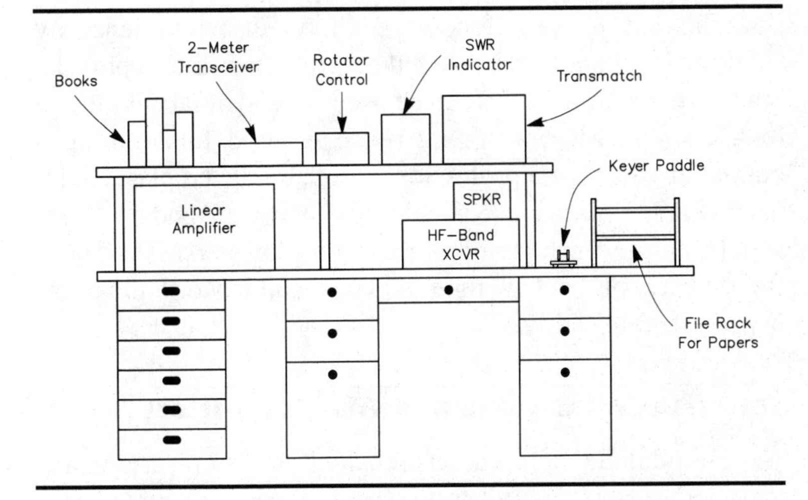

Fig 4-3—The desk layout at W1FB. The equipment is arranged for operating convenience with the most-used items directly in front of the operator for convenience. The desk follows the pattern of that shown in Fig 4-2.

Cable Arrangement

Nothing looks worse in a ham station than a tangle of wires and cables spread across the back of your desk or on the floor. Avoid this unsightly mess by cutting your cables to the exact length needed to reach from one piece of equipment to another. This includes coaxial cables from your transceiver to a linear amplifier, SWR meter and to your antenna tuner. I use RG-8X coax for connecting station equipment. It's flexible and smaller in diameter than RG-8 cable. RG-8X is approximately ¼-inch in diameter (similar to RG-59). It can safely handle full legal amateur power if the SWR is less than 2:1 between the pieces of equipment. If you're in doubt about the SWR in your station, and if you plan to use high power, use RG-8 cable or better. RG-8X is, however, less costly and contributes to a neater installation.

The ac line cords for your station can be shortened by folding excess cord back on itself several times, then taping the bundle. Kitchen twist ties can be used instead of tape if they're long enough. Meat-wrapping twine is good for securing a bundle of wire. For a professional look, ratcheted plastic wire ties in various sizes are available from hardware and electrical supply shops. This method of improving the neatness of your station takes only a few minutes. The result is worth the effort if you like things tidy.

Your Station Ground and AC Line Filters

Homemade or commercial ac line filters are worth considering as a standard, necessary station item. Your ac cords plug into a line filter (also known as a "brute-force filter") and the filter is plugged into the ac wall outlet. These filters don't work properly unless the common part of the filter circuit is connected to an effective earth ground. The ground lead should be as short as possible. A metal cold-water pipe might suffice for the filter ground.

An ac line filter helps reduce interference to TV receivers and other electronic entertainment devices. RF energy from your transmitter can migrate into the ac line and be carried directly to TV sets, hi-fi equipment and other RF-sensitive devices. This phenomenon is accentuated when you have an ineffective earth ground for your station. The RF currents flow into the ac line as they "search" for a ground return. RF energy on your ac line can follow the service lines to the pole transformer in your neighborhood. The overhead power lines may radiate like an antenna. and put unwanted RF energy near a neighbor's TV or FM receiver antenna or feed line, causing TVI or RFI. If you install an ac-line filter in your station, it will prevent RF current from passing into the power service. The filter should be as close to your transmitter as possible to

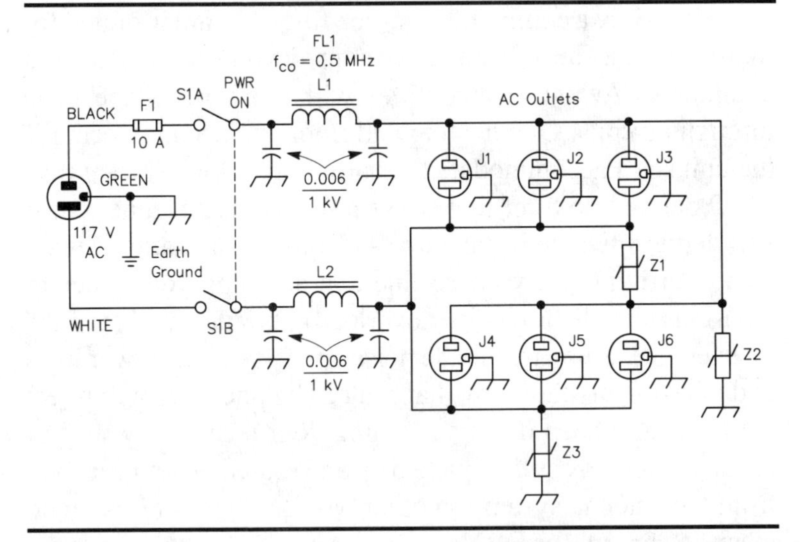

Fig 4-4—Circuit example of a brute-force ac line filter. This circuit is described in detail in December 1986 *QST* on page 25. Filters FL1 and FL2 don't allow RF energy above 500 kHz to pass into the ac line. Z1, Z2 and Z3 clip voltage spikes that appear on the ac line, protecting the equipment connected to the filter.

prevent the power cord between the rig and the filter from radiating and acting as a short antenna.

Fig 4-4 shows the circuit of a homemade line filter, presented as a construction project in *QST*. L1 and L2 are wound on ferrite rods to provide high inductance with the smallest number of wire turns. Z1, Z2 and Z3 have been added to protect the station equipment from ac-line voltage surges. These are metal-oxide varistors (MOVs), available from Radio Shack and surplus electronics dealers. The MOVs conduct when the line voltage exceeds the rated value, preventing momentary excess voltage peaks from reaching the station gear.

If you buy a commercial ac-line filter, be sure it's rated for more than the maximum ac current drawn by your station equipment. An underrated filter will cause a voltage drop through its coils. Overheating and filter damage may occur if the unit can't accommodate the current that flows through it.

Your ac line filter may serve a second useful function. If you have unwanted standard AM-broadcast station signal energy arriving via your ac line and affecting your receiver (spurious signals or receiver overload), the filter will stop the interference. Some manmade noise appears on the power lines and enters your station by that route. The line filter will reject the noise and keep it from reaching your receiver by way of the ac line. It may still be picked up and routed to your receiver from the antenna system. In other words, a line filter is by no means a cure-all for QRN.

The tiny integrated-circuit chips (ICs) and other sensitive solid-state components inside your radios, computer, TNC, etc, are particularly vulnerable to static electricity buildup. A well-stocked office supply store or mail-order catalog can help you select devices that connect to ground and help "bleed off" charges that can accumulate in your body. If you shuffle across a carpet and then reach for a button on your radio or computer, you can destroy it in one quick "pop of static electricity.

The Chair in Your Station

Hams spend a lot of time sitting at the operating desk, so you should be as comfortable as possible. Operator fatigue may result if your chair is unsuitable for long occupancy. The effects of an inferior chair will be noticed during contest operation or if you chase DX for many hours.

I don't care for hard, straight-backed chairs. I choose a swivel chair with casters and plenty of padding. It must have armrests. My second choice is a secretary's armless chair with casters and soft upholstery. I know a chap in Barbados who uses a barber's chair at his operating position. Although it looks odd in his shack, he's comfortable when he's on the air!

You can find low-cost swivel office chairs at used office-supply stores. I paid a few dollars for a nice one with a metal frame and a vinyl-plastic covering. It was dirty and needed oiling, but I cleaned it quickly with a standard household liquid cleaner. A few drops of oil stopped the squeaks. Whatever chair you select, besure it offers adequate lower-back (lumbar) support and allows you to sit upright confortably. You'll spend many hours at your main operating position, and proper posture and support will greatly improve your stamina and pleasure.

If you plan to use a swivel chair with casters on carpeting or soft linoleum floor covering, install a rigid chair pad on the surface where the chair is used. Otherwise, the casters won't turn easily and you'll probably ruin the rug. The casters will dent the linoleum if a pad isn't used. I use a clear plastic chair mat. You can make your own from tempered Masonite if you want to save money. It should be coated with two or three layers of polyurethane varnish to increase longevity.

Avoid Stacking Your Equipment

Heat is an enemy of electronic equipment. Transmitters and linear amplifiers especially need plenty of ventilation if the tubes and/or transistors are to last a long time. It's ironic that the tubes and transistors endangered by excessive heat are the devices that generate the heat. Most equipment cabinets have perforated covers and circulating fans that help to exhaust the heat that's generated. It's important to keep the ventilating orifices free of obstructions so there's unrestricted air intake and exhaust for the cabinets.

If you stack one unit atop another or allow too little space between side-by-side cabinets, the airflow will be restricted and excessive heat may develop within the cabinet. This can cause VFO frequency drift that may continue for hours.

Keep your gear a reasonable distance from the walls of your radio room. This will enhance airflow. Play it safe and allow plenty of "breathing space" for your equipment.

Summary

An orderly ham shack enhances your operating pleasure and safety. Start right by planning a neat, comfortable operating position. The reward is operating convenience and increased fun.

Chapter 5

TVI and RFI—Strange Bedfellows!

No amateur can escape the potential threat of interference to his or the neighbors' TV sets, FM receivers, hi-fi gear and other electronic gadgets. The local cable television system is subject to interference from amateur transmitters.

Hams must be skilled diplomats when working with neighbors to resolve amateur interference. It's vital for hams to conduct themselves in a courteous, understanding and helpful manner, although a neighbor may appear angry when he or she approaches you about TVI or RFI. This may seem like a tremendous responsibility for you, especially if you're a new ham, but it's a way of life you'll grow used to, especially if you live in an urban area.

Most modern electronics entertainment equipment and cordless telephones aren't designed to reject unwanted RF energy. Even though you may have a perfectly operating ham station with a pure output waveform, you can cause interference to nearby electronics equipment. It may not be your fault if there's television interference (TVI) or radio frequency interference (RFI). It remains important, however, that you attempt to correct problems that occur by offering advice and assistance to affected neighbors.

TVI and RFI are the most common forms of interference amateurs contend with. Some forms are easily resolved, although others may take considerable experimenting before the ailment is cured. No two cases of interference respond to the same "medication."

You may never have an interference problem. Much depends on the equipment you and your neighbor use, the output power of your transmitter, the way you tune and operate your transmitter, and the location and type of antenna you use. The conductive materials (phone lines, power lines and fences) around your home become a part of the complex interference equation.

Your First Responsibility

Make sure your amateur equipment and accessory items are hooked up in accordance with the instruction manuals. Each electrical connection should be tight and secure. Avoid poor solder joints and mediocre wire splices. Correct tune-up procedures, in accordance with the instruction booklet for your rig, are vital if you're to transmit a clean signal. Never try to push the transmitter power output beyond the specified amount; likewise with the mike gain and driving-power level. If your station is operating "cleanly" and you don't interfere with the entertainment equipment in your own home, the FCC will look favorably upon you if an inspection is necessary. The

Where to Find More RFI Information

If you need more information about interference, the ARRL offers two excellent sources: *Radio Frequency Interference: How to Find It and Fix It* (a comprehensive reference book), and the Technical Department's RFI Package.

You can probably find *Radio Frequency Interference* at your local dealer, or you can order it direct from ARRL HQ. The RFI Package is available for a large (9×12-inch) self-addressed envelope with three units of postage. Send your request to the Technical Department Secretary, and write "RFI Package" on the envelope to speed processing.

Another source of information is the FCC's *Interference Handbook*. For a copy send a 9×12-inch envelope with three units of postage to the ARRL Regulatory Information Branch. Write "FCC Interference Handbook" on the envelope.

If you need assistance with a specific interference problem, contact the ARRL Technical Information Service. The TIS staff will refer you to the information that best suits your needs.

FCC may ask to inspect your station if a neighbor files an interference complaint.

The goal of this chapter is to prepare you if you're called upon to deal with interference. You might be accused of causing TVI or RFI while you're at work or on vacation. This has happened to me and other hams I know. It can happen when a neighbor recalls that you're a ham and concludes that the interference has to be coming from your home. Although it's no longer required by the FCC, an accurately kept logbook can, in such instances, verify that you weren't on the air when the interference took place.

How to Use Your Transmitter

Modern solid-state transmitters contain output networks (filters) that match the transmitter to a 50-Ω load while providing harmonic suppression. Harmonic energy (energy that is an exact multiple of the operating frequency) is the most common cause of interference to entertainment receivers such as TVs and FM radios. An antenna tuner can help suppress harmonics in both solid-state and tube-type transmitters. Use a low-pass filter for added harmonic suppression.

Adjust your antenna tuner each time you alter your operating frequency by more than a few kHz. The greatest waveform purity occurs when the circuits in your transmitter are tuned to resonance at the operating frequency.

AC Line Filtering

RF energy from your transmitter can enter the ac line and be conducted to other rooms in your house. It may follow the power lines to other parts of your neighborhood. This stray RF energy may occur at your operating frequency or at harmonics of the operating frequency.

RF currents on the ac line originate in your transmitter and flow along the ac cord you plug into a wall outlet. What bad effects might result from RF on the ac line? First, this unwanted energy can reach your entertainment electronics equipment via the house wiring. Interference may result from the introduction of RF energy into your equipment by way of its ac cord. Second, the house wiring may radiate this stray energy and cause TVI or RFI through pickup of the energy by indoor antennas on TV and FM sets.

There's a reverse condition that can be caused by RF energy or noise pulses on the household ac service: Some entertainment gear generates pulses and RF energy that can follow the house wiring and enter your receiver via the ac line.

This can cause interference (QRN) when you're receiving signals. The ac line is a two-way street.

To correct this problem, install an ac line filter. You'll hear this device referred to by some hams as a "brute force filter." It contains coils and capacitors that suppress noise and RF energy that flows on the ac line. The filter should be as close to your transmitter, receiver or transceiver as possible. The shorter the ac cord between the ham equipment and the filter, the more effective the filter action. AC line filters are available commercially from electronics surplus dealers and distributors of new equipment.

Filter Selection and Ratings

When you select your line filter, be certain it's rated for the ac current (amperage) drawn by your equipment. I prefer a filter that can accommodate at least 1.5 times the current taken by the equipment. For example, if your equipment draws a combined ac current of 10 A, use a filter with a 15-A rating. A marginal filter can cause an internal voltage drop, causing your equipment to operate at reduced line voltage. An underrated filter may become hot or damaged from excessive current flow through the coils.

You don't need a separate line filter for each piece of station gear. You can use a single filter to take care of all of your equipment, provided it can handle the combined current. Keep all power cords short and place the filter as close to the array of equipment as you can.

If you use a linear amplifier, it's likely that you'll have it connected to the 240-V ac service. The amplifier does, therefore, require a separate line filter—rated for 240 V and the current taken by the amplifier. Refer to Fig 4-4 in Chapter 4 to examine the circuit of an ac line filter.

Antenna Precautions

Your transmitter may be operating perfectly, but you discover that there's a problem with TVI or RFI. Perhaps the interference wasn't present when you first went on the air, but the problem started a few days or a week ago. Where should you look first for the cause of this interference? Your first step is to examine the antenna system. Inspect the entire antenna to learn whether the solder joints and mechanical unions are clean and secure. Look for corroded joints in particular. If you observe evidence of oxidation, clean the dirty connections and resecure them. Defective joints in your antenna system can act like diodes and rectify the RF energy from your transmitter, causing harmonic currents to develop and be radiated. A common clue that indicates a poor antenna joint is erratic SWR readings when transmitting. On the other hand, the SWR may change suddenly and remain higher than normal when a defective joint develops.

You may find a coaxial cable connector that's not screwed tightly to your transceiver, antenna tuner or linear amplifier. A loose connector can act as a rectifier diode and cause harmonics to be radiated. Check the station-ground connections during your investigation. They need to be firm, positive bonds to avoid unwanted diode junctions. Your VHF transceiver (powered or unpowered) or scanner is another possible interference source. Other out-of-shack sources of interference include your TV set or a nearby TV preamp.

A rectifying antenna joint can cause interference in reverse. Specifically, you may suddenly hear voices, music and mysterious "blurps" in your receiver. RF energy from nearby commercial radio services can be rectified in your antenna system and cause harmonic energy to enter your receiver.

Transmitter Low-Pass Filter

Your transmitter may have a clean bill of health with respect to harmonic output energy, but there's a crosshatch or a herringbone pattern on the TV screen when you transmit. This is a tip-off that harmonic energy is causing TVI. This condition is most likely to occur if you live in a "fringe area," where the signal from the TV station is weak, and you don't have cable. In this situation the harmonics from your rig may be as intense or greater than that of the TV station at your site. The harmonic energy can override the TV signal and appear on the TV receiver screen.

What should you do to correct this form of interference? The standard approach is to install a low-pass filter at the RF-output connector of your transmitter or amplifier. These filters allow all HF energy to pass to your antenna, but reject energy above 35-40 MHz. Harmonics above 30-40 MHz fall into the VHF region, giving rise to RFI and TVI in FM and TV receivers. Therefore, a low-pass filter has little effect on your 160-10 meter operation. It's wise to include a low-pass filter as a standard accessory in your HF (below 30 MHz) station. VHF and UHF operators sometimes solve problems like this with high-pass or band-pass filters. As the saying goes, "It's better to be safe than sorry!"

TV High-Pass Filter

A common form of TVI is called "fundamental overload." This kind of TV interference isn't caused by harmonics of your transmitter, but the transmitter fundamental (desired) frequency (40, 80 meters or whatever) overwhelms or swamps the TV or FM receiver front end circuit. This blanks out the TV picture or FM sound. It may also blot out the sound in a TV receiver. The effect is often the same, irrespective of the VHF TV channel to which the set is tuned.

We're fortunate to have a relatively simple solution for this problem: A high-pass filter can cure the malady much of the time. All we need do is install a high-pass filter at the TV set. The term "high pass" signifies that FM and TV signal energy can pass through the filter with minor attenuation. Loud HF signals from your transmitter can't pass through the filter. This keeps your HF energy out of the FM or TV receiver front end, even though it's present on the FM or TV antenna and feed line. Install your high-pass filter on the back panel, as close to the FM- or TV-tuner antenna terminals as practicable.

Sometimes a brute-force ac line filter or common-mode choke (see Fig 1) may be necessary at the TV receiver, FM receiver or VCR to help eliminate interference. There are times when unwanted RF energy can enter the affected receiver via the ac line or via speaker wires, cables between a TV set and a VCR, etc. Get a copy of the ARRL publication entitled *Radio Frequency Interference: How to Find it and Fix It*. This is a valuable text for dealing with all kinds of interference problems.

The VCR Monster!

Video cassette recorders (VCRs) are probably the most susceptible of consumer electronics gadgets with respect to stray RF energy. Many an interference battle has been fought and lost by hams who tried to rid their VCR of effects from RF overload.

Any transmitter has the potential for disturbing a VCR. Most of these units are housed in a plastic cabinet, and this is an invitation to RFI. Many VCRs lack high-pass filters and other suppression measures. Therefore, the "door" is wide open to RFI from your station. I spent many hours seeking a cure for my VCR's problems. Until then, I heard frequent cries of despair from my living room! The remedies that worked for me are applicable to most VCR brands and models.

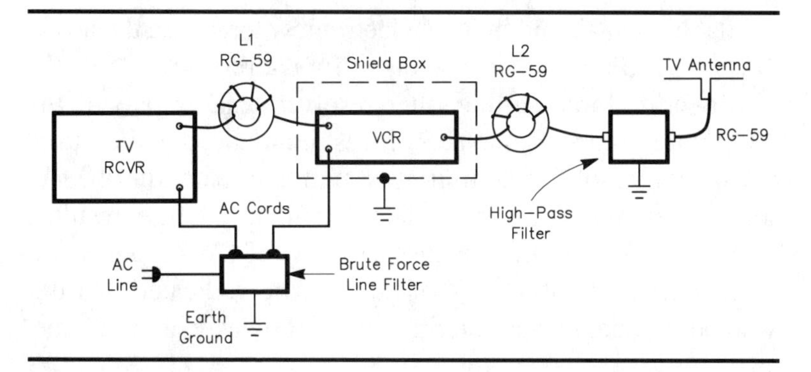

Fig 5-1—Diagram of suggested RFI suppression methods for a VCR. See text for a discussion of the devices depicted here. L1 and L2 are ferrite toroids through which a few turns of the RG-59 coax cable have been wound to form a common-mode choke. L1 and L2 should be as close to the VCR as possible.

Fig 5-1 is a block diagram of a VCR system that's been suppressed for RFI. Many steps are necessary for interference-free operation. The first is to install a high-pass TVI filter at the input of the VCR. This may not solve the RFI problem entirely or it may offer no help whatsoever. It depends on the port through which the unwanted RF energy passes and its mode of entry. In this regard, we have a "common-mode" path for the energy. In this situation, the unwanted RF energy flows along the shield braid of the coax cable and is conducted throughout the VCR ground bus to sensitive parts of the circuit. An effective suppression measure for the common-mode situation is to wrap the ac line cord and coax connection cables around a ferrite rod or loop them through a ferrite toroid core. This technique permits the desired ac and RF energy to flow through the cables without impairment to performance. These cords, when coiled around or through a magnetic core, act as RF chokes that prevent the passage of HF energy. Toroidal chokes for this purpose are shown in Fig 5-1. The core material should have a permeability of 125. Rods and toroids are available by mail and in local electronics stores (see the ads in *QST*, *CQ*, *73* and article references for addresses).

A brute-force ac-line filter resolved a large part of my VCR interference problem. RFI was pronounced while I used my linear amplifier, but it was minimal with the 100-W transceiver output power. The addition of the line filter resulted in the same level of interference at 600 W of CW output power as was observed at 100 W of power. This indicated that unwanted RF energy was entering the VCR by way of the ac line.

Telephone Interference

Cordless phones and some standard desk or wall telephones are susceptible to RFI from amateur transmitters because of design defects in the telephones. The incoming telephone lines and the cables on phones can pick up RF energy and convey it into the electronics of the phones. Solid-state devices inside the phones can act as rectifier diodes in the presence of RF energy and the resulting dc voltage can disrupt the phone's operation. Rectified RF energy during SSB operation appears in the phone as a mushy-sounding voice signal.

There are several sources of filters that can be connected to your phone for resolving this RFI problem. Your first step in seeking a cure is to contact the ARRL Technical Secretary and ask for the Telephone RFI Package.

I've been successful in suppressing phone RFI by installing 0.005-µF disc-ceramic capacitors between each active line (0.01-µF capacitors are also suitable) and the ground wire inside the phone case. These capacitors cleared up my problem 90% of the time. Stubborn RFI cases may require that you form an RF choke by wrapping the input and output phone cords around or through a ferrite rod or toroid core, as in Fig 5-1. Place these chokes as close to the phone as possible.

When dealing with cordless-phone RFI, it's helpful to add the aforementioned 0.005-µF bypass capacitors inside the cordless phone console. The hand-held portion of a cordless phone isn't as susceptible to RFI as is the main console. An ac-line filter may help to keep RF energy out of the console unit. Another source of telephone RFI is a loose or corroded fuse in the outdoor phone junction box. Also, check the box for a faulty ground-post connection. Clean all questionable joints and make sure that they have positive contacts after you clean them.

Cable TV Interference

If you have cable TV, you may never have a TVI experience with it. Or, you may have your worst example of TVI. Susceptibility to interference depends to a large measure on the type and quality of the cable converter. The integrity of the overall cable installation has a bearing on susceptibility too. You must deal with cable interference on a case-by-case basis. Rule no. 1 is to call your cable service company if TVI becomes a problem. Likewise if you receive TVI complaints from your neighbors. If your equipment is operating properly and in a clean manner, the burden of TVI suppression lies with the cable operator. It is wise to check your TV receiver, minus the cable attachments, for TVI before calling the cable company. Make certain you aren't interfering with your own TV set.

Wrap the RG-59 drop-line cable around a ferrite rod or through a toroid core, directly at the cable converter. This choke (Fig 5-1) will help to keep amateur HF and VHF energy out of the converter. Try placing the cable converter in a metal box that's grounded, if the converter is housed in a plastic case.

Don't overlook the possible benefits of using a brute-force ac line filter at the cable converter. It will keep RF energy from migrating into the converter via the ac line. You may want to use a single ac filter for your TV set, VCR and cable converter. A bonus feature of the line filter is that it will help prevent noise from the TV receiver horizontal oscillator. This reverse form of interference appears as a raucous buzz every 15.75 kHz across the tuning range of your receiver (called "TV birdies"). Most of this annoying energy is conducted to your shack along the ac line, and the ac line radiates the energy as though it were an antenna. These TV birdies are loudest on the 160- and 80-meter bands.

Neighborhood Diplomacy

Avoid an adversarial reaction if a neighbor knocks on your door or phones you about an interference problem. Gnashing your teeth and arching your back in a defensive manner will only complicate a potentially fragile relationship when TVI and RFI occurs. It's the nature of some amateurs to be defensive when a complaint is registered. After all, aren't licensed hams skilled, experienced technicians? How could you possibly cause TVI? An abrasive attitude on your behalf can cause ill will that may last forever. Furthermore, even an uninformed neighbor deserves better treatment than that.

Ask your neighbor if you can observe the interference in his home while you have a ham friend operate your station. If your equipment is causing a problem, offer your help toward solving the crisis. For example, ask permission to install a high-pass and/or ac line filter. Avoid opening his TV receiver or other consumer electronics gadget that may be affected by your RF energy.

If you're unwilling to cooperate with your neighbor, or vice versa, you may find yourself dealing with the FCC. A complaint can be filed against you through the FCC. A cooperative demeanor on your part will usually prevent FCC involvement. You may want to suggest to your neighbor that he consult his equipment dealer or manufacturer about interference that can't be corrected through conventional steps. In this instance, it's often helpful to demonstrate to your neighbor that your ham station doesn't interfere with your own consumer electronics gear.

QRP versus Interference

Above are direct methods for dealing with TVI and RFI. This chapter would be incomplete if it didn't consider a more obvious remedy for interference that affects entertainment and other consumer electronics. The greater your transmitter output power, the higher the potential for RFI and TVI. You may find that your transceiver causes no interference, whereas your station causes all manner of interference if a linear amplifier is turned on.

A practical solution for interference that's power related is to QRP (reduce power) and avoid QRO (increase power). You can reduce your transmitter power when using a transceiver by adjusting the drive control until you obtain a power-output level that doesn't cause interference. Adjust the output power while you observe the affected consumer electronics device. If you're operating in the SSB mode, reduce the mike-gain control setting until you no longer cause interference. You may not have the loudest signal in the band of your choice, but chances are that the other stations will still be able to copy your signal solidly. You may choose to operate QRP only during the hours when TV viewing and VCR use is at its peak (evenings). If you disrupt the operation of only your own household electronics gadgets, you may elect to increase the output power during the daytime. I suggest an arbitrary 25- to 50-W output power level for periods when you wish to QRP. This is ample power for most communications if you erect antennas that are effective. Many hams voluntarily observe "quiet hours," which means they avoid going on the air during prime-time periods when people watch TV or listen to hi-fi equipment. If low-power operating appears to be the only practical solution in your case, you can learn valuable operating tips and secrets by reading *Your QRP Companion*, published by the ARRL.

Summary

Some radio clubs have TVI-RFI volunteer teams. These groups will come to your house and assist in clearing up interference in your neighborhood. The team will interface with your neighbors, which can help you avoid one-on-one confrontations for you, if you have a particularly "testy" neighbor. Because the club's team is an outside entity, it's usually more successful in dealing with difficult neighbors because of its detached role in your neighborhood.

Know your local ARRL field appointees who can lend a hand with interference problems. Find out from your Section Manager or ARRL HQ who your local Technical Coordinator (TC) is. These appointees are hams who have technical expertise. They can help you resolve most interference problems (see Chapter 3 for information about TCs).

Glossary

AC line filter—A device consisting of coils and capacitors that allows the 60-Hz power-line frequency to pass through it, but blocks the flow of RF currents above approximately 500 kHz. Also referred to as a "brute-force line filter."

Antenna tuner—A coil and capacitor network used for matching unlike impedances between a transmitter and an antenna system. Also called a tuner, antenna matcher, antenna tuning unit (ATU) or Transmatch.

Brute-force filter—See ac line filter.

CATV—Cable television for use by private or commercial subscribers. (Also, an older acronym for Community Antenna Television, a system in which one well-located antenna site feeds signals to the many homes in a community where many dwellings are unable to receive useful signals on their own private antennas.)

Crosshatch —A pattern of lines that appears on the screen of a TV receiver when RF interference occurs. Generally a group of horizontal lines or bars of considerable width.

Diode joint—A poorly conductive solder or mechanical joint in an antenna or other electrical circuit. Usually caused by looseness and corrosion of the joint, causing a semiconductor junction.

Drop line—The coaxial service-entry cable that feeds CATV signals to a house from the main CATV line outside the house.

Fundamental overload—A condition that occurs in the front end of a TV or other receiver when an interfering RF signal is strong enough to lock up or desensitize the receiver-input circuit. Caused by the fundamental frequency of an amateur transmitter, rather than harmonics of that signal.

Glossary (cont.)

Herringbone pattern—Fine lines that may crisscross and shift in pattern on the screen of a TV receiver when harmonic energy (VHF) from a transmitter causes TVI.

Low-pass filter—A circuit of coils and capacitors that allows HF frequencies up to about 40 MHz to pass through it, while rejecting or suppressing frequencies above 40 MHz. Used between the transmitter output jack and the coaxial feed line to the antenna or antenna tuner.

Magnetic core—A ferrite or powdered-iron core that has permeability. It increases the inductance of a coil that surrounds it (referenced to an air-wound coil).

QRN—Atmospheric or man-made noise or static that impairs reception.

QRO—International CW Q signal that means "Increase power." When sent as a question, it means, "Shall I increase my power?"

QRP—International CW Q signal that means "Reduce power." When sent as a question, it means, "Shall I decrease my power?"

RFI—Radio-frequency interference. A condition wherein a radio transmitter's signals unintentional disturb the operation of an electronic device, such as a television set, VCR, telephone, hi-fi system, etc.

Reverse RFI—A condition wherein a TV receiver or other consumer electronics product emits energy that adversely affects reception for amateur operators, such as the horizontal-oscillator energy (15.75 kHz) from a TV receiver.

Self-oscillation —A situation wherein an amplifier stage in an audio or RF circuit becomes unstable and functions as on oscillator. This condition may occur in a dc amplifier or voltage regulator.

Glossary (cont.)

Stray rectification—Phenomenon that occurs when a defective electrical joint acts as a semiconductor. This results in RF energy rich in harmonic currents.

SWR—The standing-wave ratio in an antenna or other RF circuit. A perfect impedance match results in a 1:1 standing-wave ratio, or an SWR of 1. The ratio becomes higher as the magnitude of the mismatch increases.

Transmatch —See "antenna tuner."

TC—Technical-coordinator. An ARRL field appointee with technical expertise. He assists amateurs in his region who have technical problems. See page 8 of *QST* for the name of your ARRL Section Manager (SM), who can provide the name and address of your TC.

TVI—Television interference.

Chapter 6

Overcoming Operating Problems and Fears

All newly licensed amateurs encounter situations that can create frustration, fear or even embarrassment when you make those first on-the-air attempts. If your first time on the air is when you finish reading the instructions for a hand-held VHF FM transceiver, you may suddenly find yourself frozen, trembling and wondering how to initiate a contact on the local repeater. You may wonder, "What if I double with someone?" "What if I don't get the other ham's name right?" or "What if I time out the machine?" If you start by using CW, you may wonder, "What if I copy the other person's call sign wrong?" You may also ask yourself, "What will I do if I can't copy the other person's CW sending?"

Fear not—everyone's first time on the air is strange, shaky and perhaps a bit scary. It'll be easier for you if you remember one key point: Every ham in the world once made that nerve-wracking first contact. And most of them seem to have survived it okay!

You'll do well on FM, CW or SSB by listening first. Try a "practice" QSO in your head. Rehearse by pretending to be in on a QSO you hear on the air. If you have a friend who's a ham or prospective ham, try a few off-the-air contacts to get the rhythm together. Most important, don't set unnecessarily lofty goals for yourself. You won't garner an instant A-1 Operator nomination during your first two-way radio contact, but you are an *amateur* and it's okay to make a boo-boo. Everyone else on the band is human and has goofed a few times, so there's no reason to think you'll be any different. Join the club—get on the air and mess things up occasionally with the rest of us!

If you can't copy another station on CW, it could be that she's sending too fast for your copy speed or that she has a bad fist (poor formation of letters and words, frequent errors and no spacing between letters and words). The term "fist" is commonly used among CW operators to describe your sending ability. For example, "Wow, W1AW sure has a beautiful fist!" or "Joe's fist is so terrible he sounds like he's sending with his left foot."

Other on-the-air events may upset you until you gain operating experience. I have yet to run across a perfect operator, and that includes me! No ideal operator exists. We all make mistakes. There's a common ham-shack phenomenon called "cockpit trouble." An example of this is when you change bands and forget to switch antennas, or you operate your transmitter with the receiver volume all the way down and wonder why no one seems to be answering! At times like these, other hams may say something like, "I think I hear a station calling, but I can't copy anything," or "You have a strong signal, OM, but it's badly distorted." These things can happen if you forget one simple thing. The other operator will understand the problem because he's made the same mistakes himself.

Learning to cope with our errors is part of the big picture of Amateur Radio. Don't be remorseful if you forget a basic operating procedure or fail to copy the other station. Consider such events normal, then move ahead to your next QSO.

I vividly remember my first few contacts. I had never met a ham operator prior to becoming licensed, nor had I listened to a CW QSO. I had my Novice ticket in hand and had to muster the courage to fire up my homemade 25-W CW transmitter. My first CQ netted another Novice, but he was experienced. He used all of the standard CW abbreviations, such as HW for "how do you copy?" ES for "and" and many other shortcut terms. Each time he sent NW, I thought he meant "northwest" instead of "now." I remember remarking to my wife, "This guy is mighty strange! He keeps saying 'northwest' to me and I don't know why!" The responsibility for the problem was mine, because I didn't take the time to learn the abbreviations used by CW operators. In retrospect, I'm glad I didn't exhibit my lack of knowledge by asking him, "Why do you keep saying 'northwest'?"

Another humbling experience that occurred during my first week of Novice operating was when a General-class ham answered one of my CQs. After I completed my first transmission to him, he came back and sent QLF? I didn't know what that meant, so I went about sending my second transmission and finally completed the QSO. I learned some weeks later that QLF means "Are you sending with your left foot?" My sending with that WW II-surplus J-38 hand key must have been dreadful! I started practicing daily after that, and eventually my fist became more acceptable.

Let's discuss typical situations that can cause confusion and apprehension when you operate your station. You'll lose any fears you may have when you learn how to deal with common beginner's problems.

Inability to Copy the Other Station

We'll set the scene by assuming you've called CQ and obtained a reply from a moderately loud station (RST 589). This person is running all the letters together so that you can't separate the words and make sense of the text, and he's sending too fast for you to copy. You get upset because you don't know how to deal with this situation when he turns it over to you for your transmission. Perhaps you're tempted to shut down your station and pretend you never got involved with him. This isn't an uncommon reaction! I don't recommend that you do that, though. Instead, start your transmission by asking him to QRS (slow down). It's a normal thing to ask another operator to PSE QRS. Or you might send PSE QRS TO MY SENDING SPEED. If he slows down for you, but you're still unable to copy his sending because there are no spaces between his letters and words, be forthright and ask him to please insert some spaces between his words. Chances are that he'll try to do that and he may learn of his sending problem in the process.

Be honest and polite when dealing with this kind of problem. You have nothing to lose by asking for the cooperation of another station when you can't copy his sending. There's no cause for you to be embarrassed. A good rule for any ham to follow is to reply to a CQ at approximately the same speed as the station that calls a CQ. He may ask you to increase speed after you commence your QSO. On the other hand, it's rude to answer another station at a much higher speed than the operator is sending.

Assume that the station in the foregoing example did slow down and insert the correct spacing between letters and words. Perhaps his "guesstimate" when attempting to match your sending speed was poor. He's still sending too fast for your copying ability. More experienced hams who want to give newcomers a break often move into the Novice subbands to patiently answer CQs from newcomers and hams going along at slow speeds. This is an honorable and welcome practice because veteran operators can be very helpful on the air. Unfortunately, if they're used to sending at higher speeds or have developed their own bad sending habits, it may not be as helpful as it could be. You may be reluctant to repeat your request to QRS. Your best alternative is to copy whatever you can on paper. Even if you miss some letters and words, you should be able to piece together enough text to comprehend what he said to you. If you miss his name, QTH or your signal report, don't be bashful about asking him to repeat the key information. Always be truthful, rather than blaming your problem on QRN, QRM or QSB (fading). There's no need for you to be fearful or ashamed when asking for a repeat.

These recommendations also apply to voice operation. Suppose you contact a person who has a regional or foreign accent. His English may be limited and his accent may render some words impossible to comprehend. You're most likely to run into this situation when working DX on SSB. No one will criticize you if you ask for a repeat of data that you missed. If, for example, you have trouble getting his QTH right, ask him to give it phonetically.

Honesty in Signal Reporting

CW copy can be impaired if the signal is clicky or chirpy. The standard RST reporting system provides for letting the other person know that his signal has a defect. Many operators are hesitant about mentioning signal faults for fear of offending the other operator. But the operator with the poor-quality signal may not be aware that he has a problem. Reporting the anomaly gives him the occasion to check his signal and make necessary corrections. For example, if the other station is transmitting a loud, but clicky signal, your report would be something like RST 599K, with the K signifying clicks (or "klix"). If his signal is chirpy, your report is RST 599C.

The other ham's signal may have hum or buzz on it. If so, give him, for example, an RST 596 report. The last number in the report indicates the severity of the hum. The worse it is, the lower the last number. You may follow up this report by saying that you detect a fairly loud hum on his signal and that it sounds like 60- or 120-Hz hum.

Similar situations may arise when you're operating a voice mode. The other operator's signal may be so distorted that you can't understand what he's saying. Common causes for this malady are too high a mike gain setting, too much speech processing or a combination of both. Some operators adjust the mike gain too high in the false belief that they'll be heard better at some distant point. In a like manner, they may operate their speech processor at too high a level. I've noticed reluctance by some phone operators to tell another operator that he has bad audio. You may be doing him a favor when calling the problem to his attention. In fact, you may offer to work with him on the air as he adjusts the controls until he sounds proper to you. Make sure you have a proper voice signal, too. Follow the adjustment procedure called for in your equipment manual. I don't recommend that anyone use a speech processor if he or she doesn't have an oscilloscope or other means of observing the results of setting the levels and other controls properly.

Most splatter and distortion on the ham bands is the result of excessive mike gain and speech processing. It's against FCC regulations to transmit too broad a signal, and broad signals often can be related directly to improper adjustment of the mike and processor levels. Modern transceivers provide for monitoring the amount of compression (in dB) for the built-in speech processor. If you decide to use processing, use it only when the other station is having trouble reading you because of noise or when your signal is weak. Under these conditions, a few dB of audio compression may enable him to copy your signal solidly. I don't recommend that you use more than 3-6 dB of compression. If you exceed that limit, your signal will take on a tinny, hollow sound and may become distorted.

Some manufacturers recommend a maximum compression level of 10-12 dB. I have yet to hear a quality voice signal when the operator used that amount of compression. Misuse of the mike gain and compressors can cause a signal to be 20-30-kHz wide in a ham band. The excessive bandwidth is caused by distortion products and these products cause interference to others near your operating frequency. The amateur service is known for its self-regulating policy. You're helping to uphold this credo if you keep your signal clean, and when you courteously advise others that their signals are wide or distorted.

Don't be upset if someone tells you that your signal is weak. There will be times when you can copy the other station RST 599 or S9 plus 20 dB, but *your* report comes back as RST 549 or S6. This doesn't necessarily mean that you have a problem. There are many causes for this condition. The other person may be transmitting with an antenna that's large and high above ground, but his receiving antenna may be a short piece of wire close to ground. He may be using a kW of power, while you're using 25 W. This would explain the difference, so don't worry about the poor report. If things are so bad that he can't copy your signal well enough to continue, simply thank him for the effort and sign off. You can then look for someone who's able to copy your signals solidly.

Some amateurs believe that giving a DX station a better signal report than he deserves will help to ensure a QSL card from him. This is a case of "flattery will get you nowhere." Don't succumb to this desire, should it tempt you. Chances are that the DX station wants to know how loud his signal is so that he can better judge propagation conditions and the performance of his station. You may be doing a disservice by inflating the signal report you give him. This same rule should apply in any QSO you have.

On-the-Air Topics

I've met hams who never went on the air because they "didn't know what to talk about." They were afraid that others might find them uninteresting if they didn't use language that reflected experience or technical savvy. Don't let these feelings prevent you from pitching in and having fun. Most seasoned amateurs are happy to welcome a newcomer and they usually congratulate new hams when they learn that they were licensed recently.

There are countless nonamateur topics discussed in Amateur Radio. Avoid what I call the "form-letter QSO." This is boring to most hams. It goes something like this: "N8HLE, this is W1FB. Thanks for the call. Name here is Doug, Doug. QTH is Luther, Michigan, Luther, Michigan. My rig is a Crashboomer 112, running 100 W. The antenna is a dipole up 40 feet, so what say?" A much better first transmission, and one that would encourage a longer QSO, might go something like this: "N8HLE from W1FB. Thanks for coming back to me, OM. You have a nice signal at Luther, Michigan. Your signal is peaking at 20 dB over S9. My name is Doug and I just came in from mowing the grass. I wanted to rest for a few minutes, so I turned on my rig and heard you calling CQ. It's a pleasure to meet you. How is my signal at your location? N8HLE, this is W1FB."

This type of conversation is typical of a face-to-face discussion we might initiate when meeting another person. The form-letter QSO is cold and impersonal, and it doesn't make you sound like an interesting person. I never mention my equipment or antenna unless the other person asks me for that information. Most hams aren't interested in the details of your station unless you have something unique or if you have a tremendously loud signal. The form-letter QSO causes you to run out of conversation quickly! You may end up with an empty feeling when you terminate a QSO if you subscribe to that format while operating your station. Often the other station will suddenly get the "chow call" or tell you he has a phone call. This may indicate his boredom with the conversation. Try to be outgoing and interesting. Weather reports tend to be a crutch many hams use to get a conversation going, but they're generally meaningless to a ham many miles away. Try to exclude that topic from your QSO, unless there's severe weather at the time of your QSO that may be of interest to the other person.

Ask the other operator about his town, his other hobbies or how he got into Amateur Radio. Some hams describe their line of work and that's always an interesting subject. You need not discuss technical matters to have a good QSO. Many of us talk about fishing, hunting, computers, military experiences or some other subject. Some amateurs like to give their age during a QSO. This isn't necessary, but it's all right if you want to do it. Like any conversation, an interesting QSO may last a long time. You may even qualify for the ARRL Ragchewer's Club (RCC). All you need do is carry on a continuous QSO for 30 minutes or longer. If you want to obtain the RCC certificate, report your QSO to ARRL HQ, include a business-sized self-addressed, stamped envelope (SASE) and you'll soon receive your certificate.

One final suggestion before we move on to a new subject. Avoid such terms as "we," "QRN," "QRM" and "HI" during a voice contact. It makes little sense to say "we" when you're the only person in the ham shack. Using the term "we" may be a byproduct of the commercial broadcasting world, where the announcer is representing the owners and staff of the station. It may also indicate a kind of false modesty among people who feel uncomfortable talking about "I," "me" and "my." It's awkward to say "we," however, when it's clear that the person saying it is referring to himself alone. It's correct to say, "I'm going to work after I sign off," but it doesn't sound right to say, "We're going to work after we sign off."

Too many hams say "HI" during voice QSOs. That's the CW expression for a laugh (HI is probably a contracted form of HEE, DADADADA DIDIT evolving from DADADADA DIT DIT) and it, or HA, is entirely appropriate for Morse code. If you feel compelled to laugh during a voice contact, simply chuckle or laugh on the air. It sounds phony to say, "The joke you just told was funny, HI HI." Would you do that in person? Omit the "HI" and convey the same message.

The same advice applies to using Q-signals in a phone conversation. If you're experiencing atmospheric-noise interference, it's better to say something like, "The static is bad today," rather than, "There's a lot of QRN." In the interest of shortening the length of a sentence while operating CW, it's appropriate to send THE QRN IS BAD TODAY. Likewise for the term "QRM." Just say, "There's another station near our frequency that's causing interference." There may be exceptions if terrible conditions make it very hard to be understood or if you're talking to a foreign ham who has a limited command of English. If you follow these guidelines, you'll sound more natural and interesting to the other operator in the voice QSO.

Getting on Frequency

Are you worried about being on the proper frequency when you answer a CQ or check into a net? You may have had people tell you that your transmitter was 200 Hz high or 300 Hz low, or some such thing. This is common and it's not something that should bother you. The digital readout or analog dial on your transceiver may indicate, for example, that your signal is on 3855.00 kHz when it may actually be on 3855.200 kHz. The frequency counter in your rig may not be calibrated accurately. This means don't rely entirely on what the readout tells you. You can inspect the accuracy by tuning to WWV, adjusting for a zero-beat condition and checking the transceiver frequency readout. Most transceivers have an adjustment trimmer to calibrate the dial or digital readout. Some rigs with dials have a slip-adjustment skirt that you can change for calibration. Check your operating booklet for the calibration procedure. Make sure your RIT (receiver incremental tuning) or XIT (transmitter incremental tuning) are in the OFF position if someone says you're off frequency. This common form of cockpit trouble affects most stations at one time or another.

Consider the appropriate CW audio pitch you should tune for when preparing to answer someone calling CQ. Generally, a beat note of 700-800 Hz is satisfactory. Modern transceivers have a 700- or 800-Hz frequency offset between the transmit and receive modes, so that frequency range is a good target for you to aim for when selecting the correct received-signal pitch. Some amateurs prefer a beat note that's lower in frequency. I know hams who prefer a 100-Hz note. If you feel more comfortable with a note higher than 800 Hz or lower than 700 Hz, use your RIT to obtain that note, after you establish your QSO with the other station. Don't worry if you can't judge tones well enough to find the 700-800 Hz range. Most operators will find your signal, even if your frequency is a few hundred Hz above or below where they're listening.

Phone operation is more critical, especially if there are more than two people in the QSO. It's best to have all of the participating stations on the same frequency. This allows everyone to chat without constantly readjusting the RIT control. If your signal is a bit too high or low for the rest of the stations, have them talk you onto the frequency by telling you when you sound normal, then use your RIT to make *them* sound natural. Some transceivers have transmit-receive frequency offsets on SSB that don't match the offsets of other brands or models of transceivers. This isn't uncommon with older equipment, but this shouldn't discourage or embarrass you.

Split-Frequency Operation

Your transceiver may have two VFOs, one VFO and several memory channels, or it may have one internal VFO and an outboard VFO. You may wonder what advantage these extra features make possible. Split-frequency operation is used to communicate with a station transmitting on one frequency and listening on another. Operators use this technique on CW and SSB so that others can hear them when many stations are calling at the same time. Split operation is used mostly by operators on DXpeditions and in rare-DX countries. In this situation, use one VFO or memory channel to listen to the DX station and use the remaining VFO for transmitting on the frequency to which the DX station operator's receiver is tuned. In situations where the frequency difference is small (less than 2 kHz), you can use your RIT or XIT for split-frequency operation.

Don't be Afraid of QRM

I've known hams who say, "I can't go on the air because I can't cope with the QRM." New operators tell me how impossible it is to find a "clear frequency." It's nearly impossible to find a truly clear frequency on popular amateur bands. But imagine that you did find a clear frequency on which to call CQ. Someone answers your CQ and a nice discussion commences. Suddenly, a loud signal appears on or near your operating frequency and blots out the signal from the station you're working. You might be tempted to utter some unsavory phrases, turn off your rig and call it quits for the day. This isn't a recommended procedure. The person you're chatting with may not be aware of the interference covering his signal, and when he stands by for you, only to find you gone, he'll be greatly disappointed. What should you do in this instance? First, wait to reply to the person at the other end of your QSO. Tell him that a nearby station is interfering with your reception and wiped out his transmission. It's possible that your signal is loud enough to override the QRM, thereby enabling him to get your message. The next step, if the QRM is still present, is to choose another frequency and move your conversation. Try to salvage a QSO affected by QRM. Use your RIT or outboard VFO when looking for a clear frequency that you can move to. Your primary VFO will remain on the original frequency so that you can coordinate your move with the other operator. For example, "Let's move up 2 kHz" (on SSB), "Shall we move to 147.45?" (in a VHF FM conversation) or TOO MUCH QRM HR, QSY UP 5? (on CW). If you keep trying to convey the details of the new frequency through the QRM, the other operator will eventually acknowledge the new frequency or vice versa.

Clear frequencies are as rare as sharks in a freshwater lake. There's always signal energy that you can hear when you select a frequency. Stations above and below your SSB operating frequencies, if only a couple of kHz removed, can cause "backwash" that you can hear, however weak it may be. Your mind can function as excellent filters for separating signals. Concentrate on only the signal of your choice when using your natural filter. It takes a while to develop this ability to mentally exclude the QRM, but you can train yourself to do it. The longer you operate your station, the better your built-in natural filter. It's easier to develop this skill when operating CW because you're dealing with distinct tones or beat notes that are more easily rejected by your mind.

Don't overlook the benefits of the IF SHIFT and WIDTH controls on your transceiver, if it's equipped with those features. The WIDTH control narrows the receiver passband and the IF SHIFT control can be used to reject part of the QRM during CW and SSB operation. Narrow IF filters are beneficial; I like a 1.8-kHz SSB filter for phone operation when the going gets rough. It makes the audio sound tinny or restricted, but the readability is okay. A 250-Hz CW filter often eliminates unwanted heterodynes that cause interference. Your notch filter is excellent for eliminating a single CW beat note that may be causing QRM. Practice using these controls until you master them.

The Break-In Operator

If you're a CW operator who has just gotten on the air with a new license, you may never have considered what will happen if the station you contact decides to use break-in operation. This can traumatize you the first time you experience it. For example, you may be enjoying a ragchew, when all at once the other operator sends something like HOWS THE WX AT YOUR QTH BK? BK means "break," or go ahead and reply. He stands by for your reply and you don't know what to do. You must answer his question now, rather than wait until you make your normal transmission. You may be embarrassed because when he asked the question, you were busy taking notes on paper for use when you have your turn to transmit. You missed the question because you were concentrating on your notes at that minute. How do you deal with this problem? I've known hams who go off the air quickly, rather than let it be known that they missed the question. This isn't polite or proper procedure. All you need to do is answer the other station and ask for a repeat of the question. Simply send SRI I MISSED YOUR QUESTION BT PSE RPT BK. You'll copy his question okay this time because you'll be concentrating on it.

"QSK" is the standard term for break-in operation. Many hams prefer to use QSK all the time when working CW. A similar trend applies to SSB VOX operation—QSOs that involve break-in operation consist of short bits of conversation, followed by BK, "go ahead" or "over." This avoids the boredom that goes with the old-fashioned monologue transmissions that last 10-20 minutes. Conversations flow as if you're talking in person to some friend, while taking advantage of the break-in mode. If you don't like using VOX for SSB work, use the push-to-talk (PTT) switch on your microphone during break-in operation. This excludes the possibility of your VOX actuating when background noises occur in your ham shack or house.

Problems with Spelling

I know two amateurs who refuse to operate CW or radioteletype (RTTY) because they have trouble spelling words correctly when on the air. Yet these people spell most words perfectly when they write a letter. This is a common problem that results from not seeing the written word formed on paper. Long words are the most difficult for poor spellers to handle when on CW. I once had a problem spelling long words correctly during CW QSOs. I learned a trick that worked nicely to rectify the problem: I envisioned the word being written on the wall. It took a few tries to master this technique, but once I did, the spelling malady vanished.

You may have difficulty with spelling in everyday life. Don't let this prevent you from using CW or RTTY. Other amateurs have the same difficulty, so you're not alone! You'll still communicate the same message with an occasional misspelled word. After all, the other operator is interested in you, not how articulate you are when pounding a code key or keyboard. No one will sign off with you if you can't spell *iguana* or *pneumonia* correctly. The other person will know what you're talking about. Many long words can be converted to abbreviations during CW operation, such as XCVR for transceiver, WX for weather, etc.

Although perfect spelling during casual CW or RTTY conversations isn't usually critical, it's important for traffic handling and emergency communications. A misspelled word in a formal radiogram can make the content confusing or meaningless. A badly spelled request for aid in an emergency could lead to disastrous consequences. If you try your hand at packet radio, you should do your best. It may surprise you to read many messages and bulletins with misspellings in them. Occasional errors are normal for humans, but when you type a text on packet, take a moment to get your spelling and grammar correct. Because packet mail isn't sent in "real time," people expect you to get it right. When composing a packet message, you can easily fix a mistake or stop to look up a tricky word in the dictionary before typing it. Having a message circulate around the state or the whole country with the word "amateur" spelled "amatuer" several times in the text doesn't reflect well on the author whose call sign appears in the "From:" field.

Keyers and Keyboards

Most old-timers started out with manual hand keys, such as the old J-38 military surplus key or similar devices. It wasn't long before they craved something more sophisticated and speedy for CW work. In bygone days, the next step up the ladder was to acquire a speed key or "bug," as they're called. These keys are purely mechanical and form "dits" automatically when the key lever was moved in one direction. The "dahs" are formed manually, as with a straight key. Today, the upgrade from a straight J-38 type of key is to a keyer paddle that operates an electronic keyer. This modern electronics device shouldn't intimidate you. Anyone, unless they're physically impaired, should have no trouble learning to use a keyer. The "dits" and "dahs" are formed electronically as you operate the paddle left and right. If the keyer includes the iambic feature, you can form some words by simply squeezing the paddle levers in a certain way. This is called "squeeze keying" and takes practice to master.

You may be afraid to go on the air with an electronic keyer, for fear that you'll make many mistakes and be embarrassed. Here are a couple of recommendations: (1) Practice using your keyer for several days, or until you master it, before using it during a QSO, and (2) Don't try to operate your keyer at speeds beyond your ability—choose a sending speed that doesn't cause you to make frequent errors. The higher speeds will come to you as you gain experience.

You may eventually reach a plateau where your dexterity versus sending speed can't be increased. You may find yourself able to copy CW at a much faster rate than you can send it properly. One option you have is to obtain a keyboard keyer (KB). This system operates like a typewriter, but it sends Morse code, rather than printing letters on paper.

You need not be a touch typist to use a KB. I have no trouble sending 55 wpm with four fingers. I use my thumb for the spacebar. I never learned to touch type. A two-finger "hunt-and-peck" typist may not exceed 20-30 wpm, but error-free CW will be generated, assuming the operator doesn't hit the wrong keys. Most personal computers can be used as keyboards if the proper program is used; don't overlook that option if you have a PC in your shack.

My decision to acquire a KB came after I grew older and began having problems with arthritis in my hands. Many older hams fit this description, and the KB option is desirable. I was limited to approximately 25 wpm without making errors while using my electronic keyer. I was reluctant to change to a KB because of "honor"—I wanted to remain a purist, if possible. The KB solved my problem quickly, however, and I've never regretted the decision. I still make a few errors (typographical), but it doesn't deter me from going on the air. I encourage you to try a KB, if that's your desire.

I discovered a method that helped me reduce sending errors with a paddle: By placing a book (about an inch thick) under my hand, in front of the paddle, sending became less tiring and the number of mistakes decreased. The recommendation was made by a W4-land amateur who was having the same problem with arm fatigue and stiff fingers. You may wish to consider trying this if you have similar problems.

Multiple Comebacks—What to Do

A new ham may be confused about procedure when he calls CQ and has two or more replies to his call. Who should you answer, and how should the problem be handled?

Should you move frequency and call CQ again? Should you turn off your equipment and run for cover? The answer to these questions is a firm "no." Try to identify at least one of the call signs in the jumble of voices or CW notes, then reply to that station and mention that you heard others calling you,

too. You may announce that you're standing by for the other stations. Collect their call signs one by one and include them in your QSO, if they wish to take part in the contact. Some may elect to pull out and leave you with one or two of the callers.

There's a possibility that you may not be able to extract a call sign from the mix of signals. Suppose that in answer to your CQ, there's a WØ, N5, KB8 and a K4 station. You caught only the KB8 station's prefix. The correct procedure is to wait until all of the callers have stood by, then say something like, "The KB8 station, please call again" or "QRZ THE KB8 STATION," and your call sign. Using this method, you can sort out all the callers, one by one. You'll find that a QSO is more interesting when you have three or more people involved. The contact will last longer and you'll have more conversation to draw from when seeking a topic for discussion. A group QSO of this kind is called a "roundtable."

How to Deal with DX

Perhaps you've avoided calling DX stations because you're afraid of the language barrier. Or maybe you aren't confident about being able to get their call signs right the first time around. These are normal thoughts for a number of new hams.

Most foreign amateurs know enough English to permit a short and meaningful QSO. The international Q code is in regular use for CW DX operations, and that's why it was conceived in the first place. It allows operators to have terse communications even if they don't speak one another's language. You may not find this type of QSO especially stimulating, but you'll be able to exchange signal reports, locations (QTH) and other pertinent data. The following format might be typical in a QSO with a Russian amateur we'll identify arbitrarily as UA3DRD. If she answers your call, she'll come back to you with something like GM OM BT TU FER CALL BT NAME ELENA, ELENA BT QTH ST PETERSBURG, ST PETERSBURG BT UR SIG 589, 589 BT RIG IS 100 W BT HW CPY OM?

She'll now turn it back to you for your transmission. You may send similar information to him, in the same format. She'll understand what you've said. On the other hand, you may run into foreign hams who speak English perfectly, and you can communicate without a language barrier. If you speak a foreign language, use it when talking to a person from the country where that tongue is used. I speak Spanish with a limited vocabulary. I enjoy talking in Spanish on CW when I contact someone from a Spanish-speaking land. I have limited success using Spanish during QSOs with Portuguese hams because the two languages are somewhat similar.

If you suspect that the DX station has trouble in the English language, avoid throwing words and terms at him that he may not understand. Don't be disappointed if he doesn't want to chew the rag. He may be unable to because of his basic English. Some topics aren't permitted in some nations. Be judicious in your choice of topics when communicating with these stations, even if they speak English fluently. If in doubt, stick to subjects that concern Amateur Radio. If you have a problem relating DX call signs to the country in which they originate, don't despair. Obtain a copy of the ARRL *DXCC Countries List* and keep it handy at your desk. A similar list may be found in the international *Callbook.*

Summary

I've tried to cover the most common situations you may find difficult to face as a new ham. You'll fare best by being forthright in your operating. Don't feel bad because you lack experience. Errors are common among the holders of all license classes. Never feel that you're the only person on the air who fouls things up on occasion. A ham who never makes a mistake or offends others is a ham who doesn't exist! Stick with the learning process and your fun will increase as you become more skilled in this wonderful pastime.

Glossary

Beat note—A single audio tone heard in a loudspeaker or headphones when the receiver BFO (beat-frequency oscillator) is offset slightly from the frequency of a steady carrier or CW signal.

$\overline{\text{BK}}$—A procedural signal (prosign) that's a standard abbreviation for "Break" when operating CW, indicating that one station is standing by for a comment from another station. Usually sent in run-together manner as though it was a single character, ie, DADIDIDIDADIDAH.

Break-in—The term used to signify break-in operation (QSK), where short bits of information are exchanged during a QSO, rather than each station making a long, single transmission. Break-in also defines the act of entering a QSO in progress.

$\overline{\text{BT}}$—A prosign that's an abbreviation for "break in the text." It's used in place of a period at the end of a sentence during CW operation. As with $\overline{\text{BK}}$, it's normally sent as a single character, ie, DADIDIDIDAH

Chirp—A "yoop" or other slight frequency shift heard on a CW signal when it's keyed. Caused by an oscillator affected by the keying of the transmitter.

Click—A loud click that may be heard upon initiation or completion of a CW character, or at the beginning and end of a CW character.

Cockpit trouble—A slang expression meaning that an operator has committed an error in his ham shack.

KB—Abbreviation for keyboard keyer—a Morse code typewriter for sending CW, ASCII or Baudot RTTY. It can also mean a keyboard connected to a personal computer.

PTT—Abbreviation for push-to-talk. Involves actuating the microphone switch or a foot switch to change from receive to transmit and vice versa.

RIT—Receiver incremental tuning. Allows the operator to vary the receive frequency of a transceiver without changing the transmit frequency.

Roundtable —A QSO composed of more than two amateurs.

VOX—Voice-operated relay circuit for use during SSB break-in operation.

XIT—Transmitter incremental tuning. Allows the operator to vary the frequency of a transmitter without changing the receive frequency.

Chapter 7

On-the-Air Conduct and Procedures

Your first two-way radio contact may be breathtaking! Many new hams end up with trembling hands and unsteady voices when someone answers their first CQ! My first contact was on CW and I needed a generous serving of hot coffee afterwards to collect my wits. A few months later, I experienced similar trauma when I made my first voice contact. Being nervous during those first QSOs is by no means unusual. Don't let apprehension keep you from putting your station on the air. Chances are that the person with whom you're communicating won't notice that you're not an experienced amateur. Your confidence will increase with each QSO.

Calling CQ

Let's take first things first and address the matter of establishing a QSO. This is generally accomplished by picking a clear frequency and calling CQ. You may need to call CQ several times before you receive an answer, but don't give up. Someone will eventually tune to your frequency, hear you call and respond.

There's a proper way to call CQ. With CW operation, it goes like this: CQ CQ CQ DE N8HLE CQ CQ CQ DE N8HLE CQ CQ CQ DE N8HLE N8HLE N8HLE K. You may call a shorter CQ if you prefer, but I find the foregoing format satisfactory. Avoid excessively long CQs and try not to send a long string of CQs without inserting your call sign at frequent intervals. Some hams send seemingly endless CQs that last for a minute or longer. This discourages others from replying, especially if they must wait a long time to hear your call sign.

This applies to voice CQs on SSB or FM simplex: Call CQ two or three times, identify your station, and stand by for replies. Wait a little while because it may take another operator a moment to grab the mike or key and call you back. If there's no response after several seconds, give another CQ. A long CQ is boring to listen to. Try to keep your CQs short and allow ample time between them. This will allow time for tuning slightly above and below your frequency (use your RIT) to answer anyone who might respond to your call up to a kHz above or below your frequency. Your purpose is to attract attention and encourage someone to reply. Think of how long it takes you to clearly understand someone else's call sign, to properly tune in to their signal and to be ready to call them back. Use that amount of time as a rule of thumb. Local VHF/UHF simplex CQs need not be long, and rarely require repetitive phonetics and so forth. Usually with FM, they hear you or they don't. Regardless of the band or mode, excessively long CQs can account for a shortage of answers to CQs that are sent.

You need to find a clear frequency before calling CQ. If you fail to do this, you'll cause interference to those who are using the frequency or one close to it. Listen for a few minutes to the "clear" frequency you've chosen. It's not uncommon to find that you can't hear one of the stations in a QSO, although the other one is loud at your location. The weak station may be transmitting on the seemingly clear frequency you've chosen. The best practice in this situation is to ask, "Is the frequency in use?" on phone or the equivalent, QRL? DE (your call sign) on CW. If you're causing interference, you'll be quickly advised of the situation by another operator using the frequency. If this happens, choose another frequency and try again. During voice operation, after you find a clear frequency, speak into your microphone and ask, "Is this frequency in use?" Include your call sign. If the frequency is clear, proceed with your CQ as follows:

"CQ CQ CQ from N8HLE CQ CQ CQ from N8HLE, November 8 Hotel Lima Echo, CQ CQ CQ, this is N8HLE calling CQ and standing by."

Table 1

ITU Recommended Standard Phonetics

A—Alfa (**AL** FAH)
B—Bravo (**BRAH** VOH)
C—Charlie (**CHAR** LEE or **SHAR** LEE)
D—Delta (**DELL** TAH)
E—Echo (**ECK** OH)
F—Foxtrot (**FOKS** TROT)
G—Golf (GOLF)
H—Hotel (HOH **TELL**)
I—India (**IN** DEE AH)
J—Juliett (**JEW** LEE ETT)
K—Kilo (**KEY** LOH)
L—Lima (**LEE** MAH)
M—Mike (MIKE)
N—November (NO **VEM** BER)
O—Oscar (**OSS** CAH)
P—Papa (PAH **PAH**)
Q—Quebec (KEH **BECK**)
R—Romeo (**ROW** ME OH)
S—Sierra (SEE **AIR** RAH)
T—Tango (**TANG** GO)
U—Uniform (**YOO** NEE FORM or **OO** NEE FORM)
V—Victor (**VIK** TAH)
W—Whiskey (**WISS** KEY)
X—X-ray (**ECKS** RAY)
Y—Yankee (**YANG** KEY)
Z—Zulu (**ZOO** LOO)

Note: The **boldfaced** symbols are emphasized.

The inclusion of phonetics during your CQ helps the replying station to identify your call sign correctly. This is especially important if there's QSB (fading) or QRN (noise) that can make copy difficult.

Some amateurs use nonstandard phonetics. This is acceptable, but I recommend that you stick to the standard (ITU) phonetic alphabet (see Table 1). This makes it easier for other hams—especially those in foreign countries—to get your call sign right the first time they hear it, because hams expect standard phonetics. If I were motivated to use anything other than the ITU phonetics (which I seldom am), I might say something like, "This is W1 Foul Ball." Some hams delight in devising cute phonetics for their call signs. The choice is yours.

Specific CQs

Perhaps on a particular day you aren't interested in having just anyone answer your CQ. For example, suppose you're trying for the ARRL Worked All States (WAS) certificate. You may still need, say, the states of Alaska and South Dakota to complete your tally. In this situation, you might prefer to call CQ (on HF CW) by sending CQ CQ CQ S DAK or CQ CQ CQ KL7. This indicates to those who hear your CQ that you want to contact someone in one of those two states. You might adopt this method when you need confirmation of a foreign country when seeking a membership certificate for the ARRL DX Century Club (for confirming contacts with 100 countries). Even if you aren't chasing certificates and awards, there will be plenty of times when you may be making a specific type of CQ call. For example, on VHF or UHF FM, you might someday find yourself calling, "CQ CQ, WA1YUA looking for directions to a gas station." Always identify your station.

Directional CQs

There may be times when you want to chat only with certain stations. For example, on HF you may want to make a contact with someone in Europe, or on 2-meter FM, you may want to reach someone with a telephone. In the former situation, when operating on CW you can call CQ by sending CQ CQ CQ EU CQ CQ CQ EU DE W1FB K. Perhaps you live on the East Coast, and on a given night, California stations are coming through on 160 meters. You may want to have QSOs with only W6s. If so, your CQ would be CQ CQ CQ W6 W6 DE W1FB K.

If you enjoy taking advantage of unusual propagation modes, such as aurora on VHF, your CQ can be structured accordingly: CQ CQ CQ AU DE N1NWO, where AU stands for aurora. Another popular communication mode is known as "long path." Under this condition you can work a distant station via the longest propagation route rather than the normal shortest one. Your call would then become CQ CQ CQ LP DE N1NWO, where LP signifies long path. It's not rude to call a directional CQ. It merely lets the other amateurs know that you're interested at that time in communicating via a specific mode or with a station in a particular geographic region.

Answering a CQ

When you wish to reply to another ham's CQ by responding when he stands by, say "WB8IMY, WB8IMY, this is W1FB calling and standing by." If you're on CW, you can answer by sending WB8IMY WB8IMY DE W1FB K. There may be others who answer the same CQ. Don't let this disturb you. Most CQers answer the loudest station. It could be yours or a station hundreds of miles from you. You have the option of waiting until he completes his QSO with the station he answered, then calling him again or you may opt to answer a CQ from a station elsewhere on the band. If you're consistently unable to obtain a reply from stations you call, chances are that your signal isn't up to par. You may need to install a better or higher antenna to improve your batting average.

Operating via Repeaters

In most areas, calling CQ through a VHF or UHF repeater isn't the way things are done. It's not illegal to call CQ when using a repeater, but if you do, in most locales, the other operators will think you're a bit odd or ill-advised. The normal procedure dictates that you press your mike button and say something like, "This is N3HMD monitoring the Podunk repeater." This indicates that you're interested in having a chat with anyone who chooses to call you. Some hams make their presence known by saying, for example, "N1LMP listening" or "N8FOW standing by." This is also an invitation to those who may be monitoring the system and wish to have a conversation. If you hear someone actuate the repeater in the foregoing manner, you may answer him by saying "N2MDQ from N1NWO. Hi, my name is Frank and I'm in Goober City. Back to you." This indicates that you want to have a QSO with him. Always give your call sign when you activate a repeater. It's improper and illegal to key up the system and say nothing. People who do this without giving their call signs are called "kerchunkers," or worse still, lids! If you're checking your transceiver or antenna, say, "N8HLE testing."

Repeaters, in keeping with the FCC rules, are equipped with time-out circuits. Generally, a transmission may continue up to three minutes before the system times out and stops transmitting. Some repeaters have an even shorter timeout interval, such as two minutes. Keep your transmissions short enough to prevent repeater time out. If not, you'll be talking into the palm of your hand, so to speak, and no one will hear what you're saying! If, for some reason, you need to make a transmission longer than three minutes (not recommended), you may release your mike button and press it again. This allows the repeater timer to reset for another three minutes. Many repeaters send out a beep or tone that lets users know when it's okay to push your mike button to restart the timing cycle. Wait for this sound before resuming your transmission.

Some repeater operators or clubs don't welcome ragchewing through their system. This is because long-winded conversations deprive others of system use. This is particularly significant in metropolitan areas where many hams use a single repeater. Someone may need to use the autopatch to make an important phone call (to report an accident, for example). If the system is tied up with a ragchew and if you don't allow a pause for breakers when you start your transmission, someone may lose their life as a consequence. Maybe there hasn't been an accident, but a ham who's passing through the area needs highway directions. Long transmissions and no pauses between transmissions can deprive the visitor of local assistance.

There will be times when you run across a friend on a repeater and you may want to ragchew with him. What alternative do you have to tying up the repeater? The best practice is to ask him to meet you on a simplex frequency (but avoid the national simplex frequency, 146.52 MHz, because this is better used to call CQ or establish a simplex contact, rather than hold a lengthy conversation). Simplex operation permits you to chat without going through the repeater. You communicate by direct point-to-point operation, just as you would with CW or SSB. Simplex operation is effective only for line-of-sight communications, so your range is usually limited to a few miles if you don't have a beam antenna. Use simplex whenever you don't need the repeater to maintain contact. Regular use of simplex for VHF/UHF FM operation is the mark of a superior amateur operator.

When to Identify

It's a good idea to identify your station at regular intervals. The FCC rules require that you ID every 10 minutes during a QSO. I suggest that at the start of a contact you give your call sign and that of the station you're working. Give his call sign and yours when you terminate the QSO. You need only give your call sign each 10 minutes during the QSO. Some amateurs tend to announce their call signs too often. For example, I've heard many hams insert their call signs and those of the station being worked, each time they stood by for the other station, even though the transmissions were at times as short as 1 minute. This wastes your time and is unnecessary.

You may want to buy or build a 10-minute electronic timer (hourglasses not recommended!) that will remind you when it's time to identify your station. Many ragchewers have these gadgets in their shacks for daily use.

The Fiber of a QSO

If you're a new amateur, you may be perplexed about subject matter for on-the-air discussions. We dealt with this matter in Chapter 6, but I want to expand on the topic. Perhaps the best advice I can offer is that you avoid the "form-letter QSO." On CW it might go something like the following:

GE OM $\overline{\text{BT}}$ UR SIG IS RST 579 579 $\overline{\text{BT}}$ NAME HR IS JOE JOE JOE $\overline{\text{BT}}$ QTH IS JONES, MI JONES, MI $\overline{\text{BT}}$ RIG IS A STONECRUSHER 140 STONECRUSHER 140 $\overline{\text{BT}}$ ANT IS A DIPOLE UP 40 FT $\overline{\text{BT}}$ WX IS CLDY CLDY AND TEMP IS 68 68 DEGREES $\overline{\text{BT}}$ SO HW NW OM?

You can see that a transmission of this kind is cut and dried, and likely to be boring to the person at the other end of the circuit. It lacks the fiber needed to interest the other person

in a chat. Your first transmission can be interesting if you start with something like the following:

GE AND TNX FER CLG $\overline{BT}$ GOLLY, UR SIG IS LOUD HR $\overline{BT}$ U MUST HAVE A GOOD ANT $\overline{BT}$ NAME IS BILL ES IM LOCATED IN THE CTR OF CT, NR NEWINGTON $\overline{BT}$ BEEN WORKING ON MY ANT TODAY ES I WONDER HW MY SIG IS DOING? $\overline{BT}$ UR SIG IS RST 599 $\overline{BT}$ BACK TO U OM K

You've provided an opening for conversation by commenting on his strong signal and by mentioning that you were working on your antenna. You didn't bore him by repeating dull information that may be of little interest to him. It's seldom necessary to repeat such information as your name, QTH and his signal report. If you have a reasonably good signal, he'll get the data if you mention it once. If not, he'll ask for a repeat. Repeating information during a transmission should be reserved for times when signals are weak or when they're being affected by QRN or QRM.

Be judicious in your use of the term "HI" during CW QSOs. There's a tendency among some hams to indicate a laugh when there's no reason for one. An example is RIG HR IS A SUPERBOOMER 99 HI HI. I once heard one ham tell another MY XYL FELL ON THE STEPS AND BROKE HER ANKLE HI HI. Such a situation is by no means humorous and the use of HI is out of character. Also, avoid using $\overline{BT}$ for stalling until a fresh thought enters your head. It's not uncommon to hear something like I EXPECT THE CHOW CALL ANY MINUTE $\overline{BT}$ $\overline{BT}$ $\overline{BT}$ $\overline{BT}$ SO 73 FER THIS TIME BILL. You should use $\overline{BT}$ only once when a break in the text is in order.

Avoid Clichés

Clichés and improper use of amateur terms will tag you as a lid (poor operator). New hams sometimes adopt clichés in an effort to sound "salty" or experienced. Most clichés just sound lame. Some overused clichés include "armchair copy here, OM," "See you further on down the log," "I'm destinated," "Rig here is barefoot," and similar expressions. Instead of saying "I'm operating barefoot," why not say "I'm not using a power amplifier"? Seeing someone "further down the log" can be said better by stating, "I'm looking forward to our next conversation."

CB clichés aren't palatable for amateur use. They mark you as a dull, unimaginative person. (And some amateurs belittle CB, although many of them got their first exposure to two-way radio on the Citizen's Band.) Such expressions as, "He's got the hammer down," "10-4, good buddy," and others are inane on the amateur bands. Be your own person—be original!

Guard against misuse of the term "73." It means "Best regards" and the term stands by itself. You'll hear people say "The very best of 73s...." The literal translation of that statement is "The very best of best regardses!" It's a difficult trap to escape from once you fall into it.

"Here and There"

Try not to develop the bad habit of using the words "here" and "there" when they aren't needed. You'll hear hams who include those words as often as four or five times in a single sentence. Example: "Your signal is good here. How is my signal there? My rig here is the same model you have there." The words here and there serve no purpose in those sentences. Dialog of this kind is boring and annoying to those who haven't developed this dreadful habit.

Joining a QSO in Progress

Imagine that you're listening to an interesting QSO. The topic of discussion is especially appealing to you and you want to join the participants. How should you proceed? Avoid saying

"Break break" or "Break break break." Multiple use of "Break" is generally reserved for indicating that an emergency exists and it takes precedent over all activity on the frequency. The preferred method for breaking into a QSO is to wait until one operator stands by for the next one. At that time, you may give your call sign or you may say "Break, this is KD4WUJ." Normally, the simple mention of your call sign is adequate when entering a conversation in progress.

Don't be upset if when you join a QSO, you're more or less ignored. It can, and does, happen to all of us. As improper as that kind of treatment is, it may occur if two members of a family are having a conversation or if you aren't part of the regular group on a particular frequency. Some hams tend to be clannish and prefer to communicate with only their regular buddies. If this happens to you, take it in stride, sign out of the QSO and move on to other activities elsewhere in the ham bands.

On the other hand, if you're welcomed into a roundtable contact, be careful to keep your transmissions short and to the point. Some hams are so enthusiastic about the prevailing conversation that they tend to "windbag" when it's their turn to talk. Excessively long transmissions may annoy the other people in the QSO, especially if they've never met you. The unknown joiner in particular should avoid dominating the conversation. Hams are more tolerant of long transmissions if they know the person guilty of this act. Voice break-in or VOX operation helps prevent this from occurring.

QSK and VOX Operation

Most transceivers are equipped for voice-operated transmitting (VOX) to permit voice sounds to actuate the send-receive relay. The VOX circuit is also used for CW, at which time it's activated by the formation of the first CW character. VOX circuits have an adjustable delay time to permit fast or slow recovery of the receiver when transmitting ceases. Some transceivers are designed to allow full QSK (break-in). The receiver is activated the moment the CW key is released. This allows the other station to reply instantly and without any loss of his message.

VOX operation is commonplace on SSB. Traffic handlers and ragchewers use VOX to expedite passing information. This replaces the boring, long-winded monologues that were the norm in years past. A number of hams cling to that method of operating and one transmission may last 10 or more minutes, during which a number of topics are covered.

You'll find VOX operation more enjoyable than going "round robin" with long transmissions. A group of hams can use VOX to create a normal conversational atmosphere, such as when several people carry on a conversation in person. By avoiding long transmissions, you won't run the risk of boring others who are listening and you won't deprive a "short timer" his turn to speak: Some amateurs may have only 5 or 10 minutes to spare for a QSO. I know hams who won't join a conversation unless the stations are using VOX. Many transceivers have an antiVOX control that prevents sounds elsewhere in the house from tripping your VOX circuit. You can also work break-in by push-to-talk (PTT). This may be done with the mike button or a foot switch. There's a disadvantage with VOX operation, in that if two or more operators attempt to speak at once, they'll be "doubling" or "tripling" with one another (talking at the same time). On SSB, this isn't so terrible because all the voices just sort of mix together. On FM, however, the carriers combine to generate an unintelligible squawk that usually means no one's transmission got through clearly. You can help to avoid this problem by pausing briefly before you reply to a comment. If you hear someone else talking, wait until she's finished, then try again. I recommend that you say "Over" or "Go ahead" when you end a short transmission. This signifies that someone else may speak without doubling with you. During CW operation, you may send $\overline{\text{BK}}$ to signify that you're standing by. Don't make comments while someone else is speaking. It's not only rude, but will identify you as a lid!

The "Ah" Syndrome

Some amateurs who use VOX develop an annoying habit of saying "ah" each time they pause before a new sentence. This is done to prevent the VOX relay from dropping out until the next word is spoken. Apparently, they don't like to hear the relay action and the output from the speaker until they're finished with the transmission. They may think that the VOX circuit will clip or chop off a part of the first word in a sentence if it cycles. This can happen, but only if the VOX circuit isn't adjusted correctly. Check your operating manual for proper VOX setup. A transmission laced with "ahs" is annoying to copy. Such a voice transmission might go like this: "I plan to…ah…work on my antenna…ah…today…ah." The use of "ah" is also prompted by a need to stall while the operator thinks of the right word to use in a sentence or while he's mentally composing his next sentence. Practice speaking clearly without using "ah" for a crutch—others will find you more enjoyable to converse with.

Taboo Language

Be proud of your station and the way you conduct your activities on the air. I recognize and honor our First Amendment Constitutional rights, but I feel that coarse language, profanity and bigotry have no place in Amateur Radio. There's nothing to be gained by offending others with profane four-letter words or religious and ethnic slurs. Shabby conduct of this kind doesn't enhance our image at home or abroad. Furthermore, children often listen to your station, many of whom are or hope to someday become licensed hams, and saying nasty things sets a poor example. Imagine bringing a relative or friend into your shack to "show off" your exciting hobby, perhaps hoping to get them interested; you tune across your favorite voice subband and a string of expletives and rude banter spews out of the speaker. How do you feel about your "wonderful hobby" now? Show respect for yourself and others. Courtesy and conduct becoming ladies and gentlemen is what Amateur Radio is all about.

An important part of our Amateur Credo is to promote goodwill. You'll fare better and earn greater esteem among your peers by "keeping it clean."

Restraint is an excellent rule with regard to the tone of your conversation during a QSO. Although each of us has the right to discuss such topics as politics, sex and religion, some points of view and statements may seriously offend others with whom we chat. You may do best by avoiding touchy subjects that could make you unpopular. Many hams avoid certain frequencies on the HF bands because of the language used and suggestive nature of the conversations conducted by a small number of people who regularly congregate there. Perhaps leaving a frequency to one group's uninterrupted use isn't the best solution, but it's wisest to avoid involvement with these few social outcasts in the interest of avoiding guilt by association.

Using a Speech Processor

Few speech processors enhance the voice-signal quality. Conversely, it's not uncommon to hear SSB signals that are badly distorted because speech processing is used incorrectly. Your voice will always sound more pleasant and natural without speech processing. If this is true, you may wonder why modern rigs are equipped with processors. Manufacturers like to offer as many additional equipment features as is practicable. It adds appeal and enhances sales potential. A well designed and correctly operated processor can be an asset during difficult band conditions, but it isn't necessary to use a processor for ordinary, day-to-day communications.

I must confess, however, that 3 or 4 dB of compression from judicious adjustment of a processor can add "presence" to your signal without degrading voice quality. It depends on the type of voice you have and the performance characteristics of your mike. If you have a soft voice with a limited frequency range, you may find that a few dB of processing will be helpful for cutting through noise and QRM. Get on the air with friends who are familiar with the sound of your voice. Have them comment about how you sound with and without speech processing. This way, you can determine the best setting for the processor's compression level. Excessive processing can cause a broad SSB signal that's offensive to amateurs near your frequency. It's also illegal to transmit a broad, distorted signal. The occupied bandwidth of your phone signal should never exceed 3 kHz, but too much speech processing can make your signal several kHz wide. I've heard many signals that were distorted to the point that I was unable to understand what the

operator was saying. This seems to happen more often during contests, when some operators let their enthusiasm get in the way of better judgment. Don't use your processor unless band conditions require a bit of extra punch for your signal to be readable. Check your equipment booklet for correct adjustment of the processor, if you elect to use it. Then run on-the-air tests with your friends until you're sure it sounds better, not worse.

Another shortcoming of processor use is the exceptional pickup of background noise in your shack. Sounds like blower noise from your rig or amplifier are magnified when you use high levels of compression because the average power of your signal is increased considerably by the processor. Likewise with sounds from other parts of your home. It's not unusual to hear a signal that registers S9 on my S meter when the operator isn't talking and it may increase to 20 dB over S9 when he speaks. The residual signal is usually caused by blower noise from his transceiver's fan when he has his processor turned on. The blower noise may not be noticeable when he turns off his processor. These signals are unpleasant to listen to. Another negative feature of processing is potential shortening of tube or transistor life. The final amplifier of your rig is working much harder (duty cycle) when heavy processing is used. This increases heating of the amplifier tubes or transistors, and heat can cause gradual deterioration—or instant damage—to the amplifier devices.

Transmitter Output Power

FCC rules specify that we may use no more power than is necessary to maintain communication. Most of today's transceivers and transmitters provide 100 to 150 W of RF output power. This is ample for most communication in Amateur Radio. In fact, 100 W is frequently a greater amount of power than we need, provided our antennas are performing well and band conditions are good.

There's an unfortunate trend among some amateurs to use high power to ensure a loud signal at all times or to "hold" a frequency so that others won't encroach upon it. Hams who are guilty of this practice may be "within hollerin' distance" of each other, where maximum legal power is definitely unnecessary. As little as 10 or 25 W of power may be more than adequate under those circumstances.

High-power operation (up to 1.5 kW) should be reserved for times when poor band conditions prevail and your signal can't be copied solidly with moderate power. As a bonus when using low power, your electric bill will be much lower.

There's nothing to be gained by trying to be a "big frog in a small pond" by using excessive RF power. In fact, you'll probably cast a more positive shadow by using low power to transmit a perfectly readable signal. To this end, there's a growing number of QRP (less than 5 watts output) operators the world over. These hams love the challenge of earning WAS or DXCC awards with micro power. Others may use QRP equipment to avoid causing TVI and RFI where they live. The lower your transmitter output power, the less potential for causing interference to entertainment devices nearby.

To demonstrate how little high power may improve your signal, try this experiment. During a QSO, reduce your output power slightly before each transmission. See how far down you can go before the other operator notices any difference. You may be surprised to find that sometimes half or even a tenth of "full power" is more than adequate to conduct a conversation.

Summary

The topics introduced in this chapter represent the common areas of concern to new amateurs. Common sense and respect for others who share the bands with you should be your operating criteria. You'll be popular and respected if your on-the-air conduct is polite and proper. If you strive to maintain a clean-sounding signal, you'll enhance your position as a quality amateur. Courtesy and technical competence make a great ham and make Amateur Radio a great service.

Chapter 8

Station Accessories—What to Buy?

Some hams over-equip their stations. It's hard to resist—if you have any loose cash, there's no end to the goodies you can buy to add features, functions and frills. This chapter is aimed at helping you avoid getting in over your head for unnecessary ham station gadgets.

The basic amateur station may consist of a transceiver, a simple antenna, a mike and perhaps a packet terminal node controller (TNC) or a code key. With this modest array, a ham can ragchew, take part in contests, handle traffic (messages) and perhaps manage to work some DX.

Live within your hobby means. Buy only items you need for routine operation. In any leisure-time pursuit, there's always a temptation to overspend. This has been my fate on more than one occasion, and the temptation still nags me when

I attend a hamfest and see all of the neat things being offered by the commercial vendors and flea-market hawkers. It can also lead to personal financial hardships and domestic turmoil. Keep perspective and don't go overboard. You'll have more fun adding to your station gradually and enjoying each piece of equipment fully until the time comes when it's appropriate to add something new or replace something old with something better.

Bugs and Paddles and Keyers...Oh, My!

If you're going to set up a CW operation, you have many choices for equipment that generates Morse code information. The basic and oldest setup is a plain straight key that plugs into a jack on your transceiver. Many classic radio fans would never use anything else. A common straight key will last a lifetime, but it limits the maximum CW sending speed you can attain. A skilled straight-key operator can send good Morse code as fast as 18-20 wpm. It's easiest to begin with this simple type of key. If, as you gain experience, you decide to make a career of CW operating, you'll want to graduate to a keying system that will permit faster sending.

Fig 1—The straight key is the basic Morse code sending device. Many new CW operators learn on a straight key they buy or build.

A "bug" style of key can be your next step toward faster operation. It's somewhat outdated by today's standards, but this mechanism will last forever if it's cared for properly. A bug key allows the dots to be formed automatically by virtue of a bouncing spring-loaded contactor. The dashes are formed manually by pressing the paddle lever in the opposite direction. A common pitfall associated with bug keys is that some operators use a poor dot-dash ratio. For example, the dots may be formed at 30 wpm, whereas the dashes are only 20 wpm. This is not good Morse code! Another quirk that develops with some bug users is a strange sending rhythm that makes copy difficult. This is called a "Lake Erie" or "banana boat" swing. Ship operators used to send CW in this manner, and hence the names. A pronounced Lake Erie swing causes the words being sent to sound like something other than what's intended. A case in point is "CQ," which may come out as "NNGT." The faster the operator sends with a swing, the more difficult it is to copy his sending. Those who copy CW with a computer and video monitor will find gibberish displayed on the monitor screen. If you use a bug key, try to send proper CW. Tape record your sending, then listen to it. If you have bad sending habits, you'll detect them quickly if you adopt this method of quality control.

Advantages of a Paddle and Keyer

The fastest CW operators use a device called a paddle. A paddle key (similar to a bug) is used with an electronic keyer. The paddle has a center-off position. When you release the paddle lever with your fingers, there's no electrical connection to the keyer unit and hence, CW characters are not sent. Dahs are formed automatically when you press the right side of the paddle lever. Dits are formed when you press the left side of the lever. With an iambic keyer, if you press the right-hand-side paddle to the left to generate a series of automatically repeating dahs, you can tap the left-hand-side paddle to the right and the

Fig 2—Paddles are used with keyers to send code electronically.

electronic keyer will generate alternating dahs and dits. With practice, you can rapidly send Morse code characters made up of dits and dahs with an iambic keyer because it generates perfectly timed and spaced dits and dahs within each character. An experienced CW operator using a paddle and iambic keyer can send 50-75 wpm by hand. The paddle actuates electronic circuits that operate solid-state switches in the keyer. These switches cause your transmitter to be keyed.

The keyer has a speed control that you can set for your preferred sending speed. Most keyers also have a weight control, which acts somewhat like a sustaining pedal on a piano. Excessive weighting can render a CW word unreadable. The higher the sending speed, the worse the condition. Use proper weighting if you want to send good CW.

Many experienced CW operators claim that if you plan to eventually become a high-speed CW expert, you should start with an iambic paddle keyer. Because a paddle lets you send much faster than a straight key, and because it can be difficult to retrain yourself to use a paddle keyer after you're used to a straight key, it can be easier to use the paddle keyer right from the start. Other amateurs believe that all hams should be proficient with a straight key. Try both kinds before you decide

Fig 3—This keyer features a variety of electronic controls and adjustments, including code speeds from 8 to 50 words per minute, adjustable weighting of dashes compared to dots, and several memories that allow messages (such as CQ CQ CQ DE W1FB**) to be sent automatically.**

which to use in your shack. You might be capable of operating skillfully with both types.

Keyboard Keyers

Keyboard keyers are electronic Morse code typewriters. A skilled typist can send perfect code with a keyboard. The keyboard is connected to a special adapter or to a computer running software that generates perfect CW characters and inserts proper letter and word spacing. Personal computers can be used as Morse code keyboards with appropriate software. If you aren't a "CW purist," you may find a keyboard to your liking.

A system like this can have a built-in buffer or memory that enables you to send ahead of what's coming out over the air. A keyer I owned could store up to 255 CW characters and had four memory channels that allowed short messages to be stored and recalled. Such text as CQ CQ CQ DE W1FB, W1FB K can be kept in the memories for frequent use. I stored QRL? DE W1FB in another memory bank. Keyboards are especially handy for contest operation. Not only can you store messages such as QSL

RST 599 TU ES GL DE W1FB, but you can catch up on your log while the buffer is being emptied of its message.

More sophisticated computer software can store unlimited text files on disk, interface with logging software, contesting programs, and can automatically "read" DX spots from *PacketCluster* stations, place the other station's call sign in your log and generate the characters for you to call the station and exchange QSO information. With graphical user interfaces, such as Microsoft Windows and IBM OS/2, a high-resolution monitor screen can display your transceiver's controls, DX maps and more. You can choose the CW message you wish to send by moving your computer's mouse or trackball and clicking a button. The latest *digital voice keyer*s let you record your own voice as computer data files and send prerecorded call signs and phrases. With the proper software, they can assemble common words and expressions from recordings of standard phonetics and even from recordings of various basic syllables.

Microphones

The human voice is a complex blend of audio-frequency waveforms and harmonics that's difficult to reproduce faithfully, especially over staticky, low-fidelity radio circuits. To wring the best intelligibility from the sounds coming out of your mouth, you need a good mike for your station. A cheap microphone with inferior frequency response can make the best transmitter sound dreadful. Being pennywise and pound foolish isn't sensible if you want to have your voice signal be clear, understandable and pleasing. The stock hand-held mikes that come with transceivers are generally adequate for casual phone operation. I've found most of them too bassy for my voice, but those who don't have deep voices will fare well with a stock mike.

A desk mike with a push-to-talk stand is more convenient for home-station use. This type of mike makes it unnecessary to pick up a hand mike each time you speak. The desk mike is better for VOX operation because the mike is in front of your face when needed. You won't have problems with noises from handling the mike, nor will you be as likely to bump it into something or drop it on the floor! The best kind for serious operation is a boom mike; the type that attaches to a pair of headphones and curves around with its element in front of your mouth. With a boom mike, you can't inadvertently turn away or move too close or too far back. You can turn your head, lean back or even get up and move around the operating position. All you need is a way to control the boom mike conveniently. This can be done with a regular momentary contact push-to-talk (PTT) switch, a locking mike switch, a foot-pedal switch or by using voice-operated (VOX) operation. (A VOX circuit uses the mike to control a switching circuit that senses when you're talking and turns the transmitter on only when you're speaking.)

Avoid using amplified mikes that aren't designed to suppress RFI. Amplified mikes sold for the Citizen's Band market are especially susceptible to RFI problems. Stray RF energy in the ham shack can enter one of these mikes and cause your signal to be a mess of howls and garbled words.

Mikes and microphone audio circuitry with features such as bass- and treble-boost circuits are described in many *QST* articles. An article in August 1989 ("Build a Low-Cost Booster Microphone") shows how to make your own high-quality mike for less than $10. The circuit is RF-suppressed and contains an adjustable audio amplifier. The boost circuits enable the operator to tailor the audio for his voice characteristics. Hams with bassy voices can roll off the low frequencies and accentuate the highs. Those with high-pitched voices can roll off the highs and boost the lows. Circuits boards for the mike are available by mail (see article footnote). Some microphones designed for use on SSB have room in the head for more than one element, so you can use the more natural "wide-range" element for casual, local ragchewing, and switch to a crisper (less bassy), punchier sound to pierce the noise on crowded bands during contests or in DX pileups.

Audio processing has come a long way with advances in semiconductor and microprocessor technology. You can buy or build a sophisticated speech processor that incorporates digital signal processing (DSP), which can be used to finely tailor your transmitted audio. Many experienced hams hear the subject of speech processing and immediately react with a reflexive "Ugh!" Be careful of speech processors, however—one of the most common problems heard on the ham bands is the large number of phone operators who don't adjust their speech processing circuits correctly. The result can be a badly distorted voice, and annoying overcompressed sound or even an illegally wide, splattery SSB signal. Take the time to make the proper adjustments to this sensitive device and test it on the air extensively with friends who can listen or record your signal and help advise you on what sounds best.

Never try to force a speech-processing device beyond the limit of its intended function. Used judiciously, it can sometimes make your voice clearer to a receiving station, but it should never be used to try to compensate for having inadequate output power or a weak signal. Hams get into more trouble when they "crank up" the processor in a misguided attempt to get through on a crowded band or when propagation conditions are bad.

Be mindful of the impedance of your mike vs the characteristic of the mike input of your rig. Most modern transceivers require a 600-Ω mike. Some older rigs call for a high-impedance mike (typically 50 kΩ). A mike of the wrong impedance can cause a loss of frequency response and reduced audio amplitude. Maximum power transfer occurs when impedances are matched.

Data Controllers

The computerized digital revolution has become a permanent part of Amateur Radio. Before 1980, most hams had the choice of a handful of analog and primitive digital modes, including CW, voice, radioteletype, slow-scan television (SSTV) and facsimile (fax). Today's hams are widely active on packet radio, and many ply the airwaves with the more exotic signals of AMTOR, PacTOR, CLOVER, G-TOR, fast-scan television (FSTV) and other communication technologies. It's possible to send a full-color photograph in digital form or a complete working computer program halfway around the world. Microprocessor chips are so powerful and small that a box that fits inside a shirt pocket can be wired to a transceiver and make it possible to exchange messages and computer files to almost any other ham with compatible equipment. Gateways transfer digital information to and from amateur satellites, VHF-to-HF links, through worldwide landline networks, and to manned spacecraft. Beginners can choose a basic controller capable of one or two digital modes, a multimode data controller or a programmable all-purpose device that can function in almost any mode by using digital signal proessing and by doing the encoding and decoding with powerful computer programs.

If you're interested in these modes, it's generally less expensive to start out with a simpler device, such as a basic packet TNC. On the other hand, if you plan to move forward and explore several digital modes, a multimode unit may prove to be a better initial investment because it costs less to buy one full-featured device than to add on to or upgrade an entry level controller.

Choosing a "plain" packet-only TNC doesn't limit you severely. Many amateurs, through their business careers or their other hobby activities, have considerable experience with computer networking over wires. If you enjoy dabbling in this

Fig 4—A terminal node controller (TNC) is used with a personal computer to operate on packet. Computer software controls the TNC. More sophisticated devices, called multimode controllers, allow operation on a number of other digital modes, such as AMTOR, PacTOR, Clover and G-TOR.

area, you might want to experiment with more advanced packet radio systems, such as the TCP/IP software based on code first developed by Phil Karn, KA9Q, in the mid-1980s. The software (usually called *NOS*, *NET* or some variation) is free and has taken off dramatically since about 1993. Its popularity has grown partly because so many amateurs now have powerful computers in their home shacks. Because it's based on data-exchange protocols accepted worldwide, hams can link their stations over the airwaves and across paths that may travel partly along the high-speed wires and fiber-optic circuits of the Internet. This is a fascinating field of ham communication and is wide open to experimentation, development and potential newsmaking breakthroughs!

Antenna Tuners

We discussed antenna tuners earlier in this book. The issue is whether you need a tuner. Few shacks are without some type

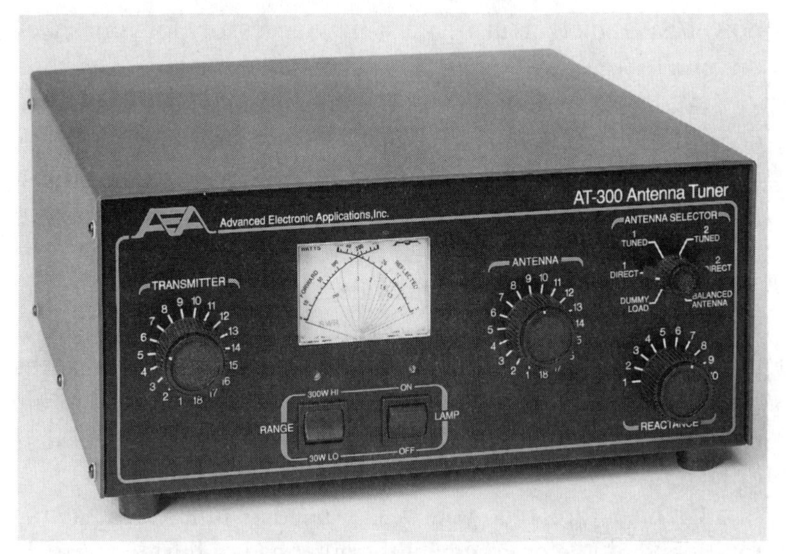

Fig 5—An outboard (separate) antenna tuner allows more precise matching than the units built-in to many modern transceivers. You may find one useful if you use a multiband wire antenna. This model also has an SWR/power meter.

of coil-capacitor network that serves as an antenna tuner (also known as an ATU, matcher, antenna coupler or Transmatch). Hams use these gadgets between the transmitter and the antenna feed line to ensure a match to the 50-Ω transmitter. Tuners are also used to match a transmitter and receiver to end-fed wire antennas. Although most hams who operate HF stations are familiar with tuners, few VHF/UHF stations are equipped with these devices. All hams should be able to adjust the match of their transmitters and antennas, though. Even if it works, operating from 144 to 148 MHz without tuning adjustments may not be as effective as you'd expect. Don't assume that your radio or antenna is able to operate at maximum efficiency across a 4-MHz range! An accurate power/SWR meter is also a must-have accessory for your shack or vehicle (see below).

A tuner is essential for use with a multiband antenna that doesn't have traps and a coaxial feeder. A multiband dipole is often fed with 300 or 450-Ω ribbon line, a 4:1 balun transformer and an antenna tuner. With this setup, you can use a 160-meter dipole for all of the HF bands up to 10 meters. The antenna tuner is adjusted to provide a 1:1 SWR at the operating frequency. In some instances, you may need to convert 2000 Ω to 50 Ω, or 30 Ω to 50 Ω. It depends on the antenna and the operating frequency. Various impedance combinations are common when using a multiband antenna with tuned feeders or an end-fed wire.

Chances are that you won't need a tuner with a trap multiband dipole or vertical if the antenna is adjusted correctly and the SWR is less than 2:1. The 1:1 SWR is better because it indicates a matched condition and maximum power transfer will occur. If you want to use your coax-fed multiband antenna in a part of a ham band where the SWR is high, you can use an antenna tuner to lower the SWR to 1:1. This doesn't correct the mismatch at the antenna feedpoint, but your transmitter will be connected to the desired 50-Ω impedance.

Antenna tuners cause power loss, as do balun transformers. These are passive circuits (no operating voltage required) and losses occur from heating and circuit resistances. Don't fall into the old trap of panicking over losing a couple of watts in the feed system. A well-designed tuner that's adjusted properly shouldn't cause more than 1 dB of power loss. Losses of 0.1-0.3 dB are typical. There's no need to worry about this loss because you'll gain dB by providing a matched condition with your tuner. Whether you need a tuner depends on the kind of antennas you use and the frequencies you use within a given amateur band.

Automatic Antenna Tuners

Computerized, automatic antenna tuners allow you to push a button and let the tuner adjust itself for a 1:1 match, or nearly so. Automatic tuners are sometimes large and expensive. They offer nothing that can't be done with a manual tuner, but they enable you to change frequencies rapidly and without playing with a couple of knobs. If you get the hang of it easily, you may want to avoid buying an automatic tuner and be content with a manual tuner that requires the adjustment of two or three controls. Automatic units are excellent for handicapped operators. These tuners are an asset for contest operators who must change bands quickly.

It's easy to build an antenna tuner. Consider constructing your own unit with components gleaned at flea markets or from electronic surplus dealers. The *ARRL Handbook* and the *ARRL Antenna Book* contain circuits for homemade antenna tuners and balun transformers.

SWR Indicators

You'll hear about SWR meters and SWR bridges. Not all SWR meters contain a bridge circuit and some are reflectometers. Both styles of indicator serve the same purpose, with regard to letting you know when your feed line is matched to your transmitter. They can also indicate when a feed line is matched to an antenna. Although some SWR indicators have only an SWR scale (Forward and Reflected), other instruments have a meter scale that registers the power output of a transmitter. The latter instrument is called an RF power meter. Modern versions of this unit may provide peak power readings for use in measuring SSB output power.

You need an SWR indicator if you plan to build antennas or use an antenna tuner. Two home-brewed SWR meters were described in July 1986 *QST* ("The SWR Twins—QRP and

Fig 6—An outboard SWR/power meter measures power output and SWR more accurately than those built-in to some transceivers.

QRO"). They're inexpensive and easy to construct if you don't want to buy a manufactured instrument.

SWR indicators and RF power meters are designed for particular impedances, generally 50 Ω. You must used them in a system that has 50-Ω transmission line. Don't use a 50-Ω SWR indicator in a 75-Ω feed line. If you do, the readings won't be accurate. You can't use a 50-Ω SWR indicator between your transmitter and an end-fed wire antenna: An antenna tuner is necessary between the wire antenna and the SWR meter to ensure a 50-Ω load at each end of the instrument.

Outboard Audio Filters

You can use active audio filters between your receiver and a loudspeaker or headphones. Audio filters contain solid-state electronic filters. Some provide only a band-pass response and fancier ones allow you to choose a high-pass, low-pass or band-pass audio response. Most hams don't need an audio filter, but one can be helpful when you try to copy a weak signal or reject QRM from a signal close to your frequency.

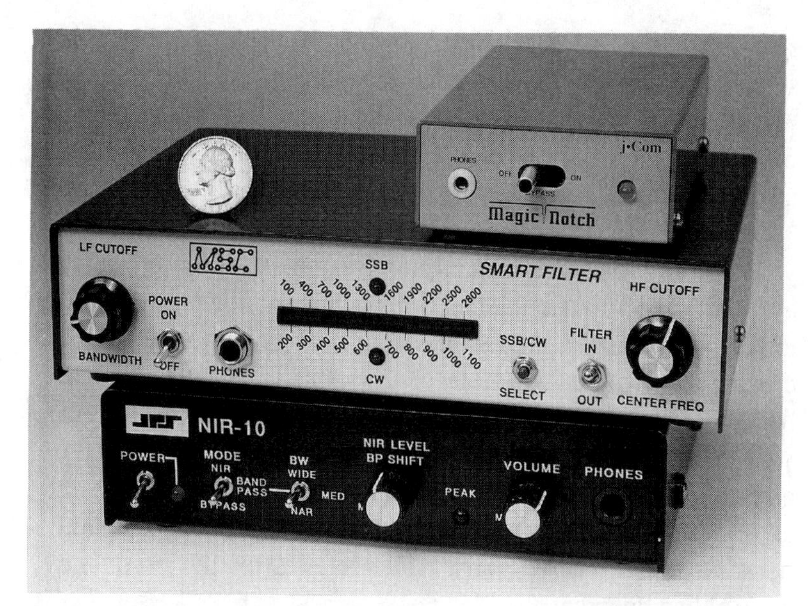

Fig 7—An audio filter such as one of those pictured here can help reduce interference from strong signals near the one you're trying to copy.

What does an audio filter do? It "launders" the signal that comes from your receiver by reducing the effects of wideband noise (developed in your receiver). It can enhance and lift a weak signal above the background noise when you operate SSB or CW. There are times when an unreadable voice or CW signal can be copied solidly while using an audio filter. I use one all the time when I'm operating CW on HF, even when loud signals are present. If your receiver has a narrow IF filter for CW work (250 or 600 Hz), however, you may not realize a significant benefit from using an audio filter. If your receiver has only an SSB-bandwidth filter, the audio filter will help considerably when you try to sort out CW signals.

Most audio filters have a Selectivity control that you can use to make the filter response narrow. The greater the selectivity, the better the rejection of QRM. Maximum filter selectivity and sharp bursts of static can cause a CW signal to "ring" (sound mechanical or tinny). This condition can impair readability and is annoying to the operator. It takes practice to adjust an audio filter for a "just-right" sound.

Most audio filters have a Frequency control. This allows you to tune the filter for maximum response (audio output) at a CW pitch of your choice. Commercial amateur transceivers are usually set up at the factory so that a correctly tuned signal creates an audible CW note of 700-800 Hz. Most amateurs have a favorite CW pitch that may differ, however, ranging from about 100-1000 Hz. The proper procedure for filter adjustment is to tune in a CW signal pitch, then turn on the audio filter and adjust maximum response. The Selectivity may be adjusted for the response you prefer, after the Frequency control has been set.

I don't find an audio filter especially helpful during SSB operation. You can adjust them to reduce the effects of some types of QRM, but they don't usually provide the good results you can obtain with a narrow (1.8 kHz) SSB IF filter used in combination with the IF Shift control on your receiver. If you don't have a narrow SSB filter, you can still reduce QRM by using your IF Width control with the IF Shift feature. Your IF width circuit allows you to narrow the response of the receiver IF circuit in a progressive manner (variable selectivity). This may approach the effectiveness of a narrow SSB IF filter.

The best choice for an all-around audio filter is a DSP device. These don't usually create the "ringing" of standard filters, are imperceptibly fast and accurate, and permit finer adjustments and additional ways to filter incoming audio signals. A DSP filter takes the audio from your receiver, samples it and analyzes it thousands or millions of times each second, and applies customized filtering and processing to the audio waveform before it's passed on to your loudspeaker or headphones. As computer component development has improved dramatically over the past few years, DSP applications in Amateur Radio have become sophisticated, easier to use, less expensive and more commonplace. It's likely that almost all amateur signals—transmitted and received—will be processed digitally in the future.

Antenna Rotators

If you plan to erect a Yagi or cubical-quad beam antennas, you'll need a husky antenna rotator to turn it. It's important to choose a rotator designed for turning large antennas. Most TV rotators can handle a small or medium-sized VHF or UHF Yagi, but won't accommodate large arrays or HF beams because ratcheting occurs when high-velocity winds prevail. The gears can be stripped in short order. The gears can also be damaged when rotation ceases and there's a momentary strain on the unit while the antenna movement ceases. TV antenna rotators are less expensive and often satisfactory for turning small VHF and UHF arrays, where weight and wind loading is minimal.

I recommend that you purchase a rotator designed for amateur service. These mechanisms have a built-in brake that prevents gear damage when the beam antenna is subjected to high velocity wind. The brake is released electrically when you rotate the antenna and engaged again after all motion stops. Don't try to save a few dollars by purchasing too light a rotator. A unit designed for the application will last for years and you'll have peace of mind when big windstorms come along.

Do You Need a Tower?

You need a tower if you anticipate erecting a beam antenna. A tower isn't necessary for supporting most wire or vertical antennas. Wooden poles, trees or telescoping steel masts are generally best for holding these simple antennas aloft. If you're a VHF/UHF operator or a DX chaser and want

Fig 8—Many experienced hams have found that nothing beats a rotatable directional antenna (or antenna *array*, as shown in the photo) for top-notch performance. If you get involved with serious contesting or DXing, you'll probably want to install a tower-mounted antenna at some point—assuming you have the necessary real estate.

your antennas to be high above the ground, a tower may be your best bet. Wooden poles, trees and telescoping masts have a practical limit with regard to height.

You can save money by purchasing a used tower. Try to find one that's not rusty or bent. If it has rusty areas, use a wire brush and sandpaper to clean the bad areas, then apply automotive primer paint to the areas you've cleaned. Follow this with a coating of outdoor aluminum paint. Consider painting the entire tower to prevent new rust spots from developing.

Tilt-over and crank-up towers are the most costly new units. They offer an advantage over stationary towers. This is especially true if you don't like to climb towers! I use a 50-foot tilt-over tower that has one set of guy wires (at the 20-foot level). It's set in four cubic yards of concrete. It's a joy to crank over the top 30 feet of tower when I need to work on my antenna or when I want to try a new antenna. The hand-operated winch enables me to tilt the top of the tower to ground level in about five minutes.

Tilt-over and crank-up towers present safety hazards. A broken lift mechanism or frayed cable can cause the tower to come crashing down while lowering or raising the tower sections. Keep these parts of the system in good repair. If the cable appears worn or rusty, replace it. The winch needs to be oiled and checked at regular intervals.

Check your local zoning ordinances or neighborhood covenants before you invest in a tower. You may live in an area where towers are prohibited. Your city engineer may be consulted for information about ordinances and installation requirements. The Regulatory Information Branch (RIB) at ARRL HQ maintains a store of literature and a tremendous resource of information and recommendations for hams who face antenna or tower restrictions. Call RIB for information.

Phone Patches

Most amateurs have no regular need for a phone patch. This is a device that allows you to connect your telephone to your amateur station. A patch enables hams and nonhams to communicate with friends and relatives with the phone service routed through your transceiver. A patch is an electronic device that's inductively coupled (not by direct connection) to the phone line in your home and to the station transceiver. Phone messages are sent over the air, picked up by another ham's station and fed back into his local phone line. Some amateurs specialize in handling phone-patch traffic for servicemen and women at distant military bases. Some amateurs use patches for their own convenience to communicate with friends and relatives who live far away.

For important tips on using telephone patch equipment, see the sidebar, "ARRL Phone Patch Guidelines."

Dummy Antenna

Dummy antennas—wrongly referred to as "dummy loads"—are an important part of a ham shack. It's effectively a "fake" antenna that provides a precise 50-Ω impedance match to the transmitter's output, but instead of radiating the signal through space like an antenna, dissipates most of it as heat, thus minimizing the chances of your tests interfering with other stations or devices. Get a 50-Ω dummy antenna for your station. They're available from ham equipment vendors or you can build one from readily available, inexpensive parts. Some dummy antennas are air cooled and others have large 50-Ω resistors immersed in cooling oil. They're equipped with a coaxial connector so that you can attach the load to your transmitter with 50-Ω coaxial cable. A dummy antenna must be capable of accommodating the full output power of your transmitter without overheating and becoming damaged. Check the manufacturer's ratings before you purchase this

ARRL Phone Patch Guidelines

The information in this section was developed to help guide amateurs who use autopatch facilities on VHF/UHF FM repeaters, but most of these rules and recommendations apply to all Amateur Radio telephone patch operations.—Ed.

Telephone patch operation involves using a ham radio station as an interface to a local telephone exchange. Hams operating mobile or portable stations are able to use a phone patch to access the telephone system and place a call. Hams use autopatches to report traffic accidents, fires and other emergencies, or to communicate with family and friends while far from home. There's no way to calculate the value of the lives and property saved by the intelligent use of phone patch facilities in emergencies. The public interest has been well served by amateurs with interconnect capabilities. As with any privilege, this one can be abused and the penalty for abuse could be the loss of the privilege for all amateurs. The suggested guidelines here are based on conventions that have been in use for years on a local or regional basis throughout the country. The ideas they represent have widespread support in the amateur community. Amateurs are urged to observe these standards carefully so our traditional freedom from government regulation may be preserved as much as possible.

1) Although it's not the intent of the FCC rules to let Amateur Radio operation be used to conduct regular, routine commercial activities, phone patching involving business affairs may be conducted on Amateur Radio. (The FCC has stated that it considers nonprofit and noncommercial organizations "businesses.") On the other hand, amateurs are strictly prohibited from accepting any form of payment for operating their ham transmitters, they may not use Amateur Radio to conduct any form of business in which they have a financial interest and they may not use Amateur Radio in a way that economically benefits their employers.

Amateurs should generally avoid using Amateur Radio for any purpose that may be perceived as abuse of the privilege. The point of allowing hams to involve themselves in "business" communication is to make it more convenient and to remove obstacles from ham operations in support of public service activities. Before this rule was revised in 1993, it was often technically illegal for amateurs to participate in many charitable and community service events because the FCC regarded any organization, commercial or noncommercial, as a business with respect to the rules, and prohibited hams from making any communication to in any way facilitate the business affairs of any party. That meant that operating a talk-in station for a local nonprofit radio club's hamfest constituted a violation!

Now it's legal to use ham frequencies, including telephone patch facilities, to communicate in such a way as to facilitate certain personal business transactions. The distinction is essentially whether the amateur operator has a financial stake in the communication. This means that a ham may use a patch to call someone about a club event or activity, to make a dentist appointment, to order a pizza or to see if a load of dry cleaning is ready to be picked up. In such situations, the ham isn't in it for the money. However, no one may use the ham bands to dispatch taxicabs or delivery vans, to send paid messages, to place a sales call to a customer, or to cover news stories for the local media (except in emergencies if no other means of communication is available). If the ham is paid for or will profit from the communication, it may not be conducted on an amateur frequency. That's why there are telephones and commercial business radio services available.

Use care in calling a business telephone via amateur phone patch. Calls may be legally made to one's office to receive or to leave messages, although using Amateur Radio to avoid the cost of public telephones, commercial cellular telephones or two-way business-band radio isn't considered appropriate to the purpose of the amateur service. Calls made in the interests of highway safety, such as for the removal of injured persons from the scene of an accident or for the removal of a disabled vehicle from a hazardous location, are clearly permitted.

A final word on business communications: Just because the FCC says that a ham can place a call involving business matters on a phone patch doesn't mean that a station licensee or control operator *must* allow you to do so. If he or she prefers to restrict all such contacts, he or she has the right to terminate your access to the system. A club, for instance, may decide that it would rather not have members order commercial goods over the repeater autopatch and may vote to forbid members from doing so. The radio station's licensee and control operator are responsible for what goes over the air and have the right to refuse anyone access to the station for any reason.

2) All interconnections must be made in accordance with telephone company rules and fee schedules (tariffs). If you have trouble obtaining information about them from telephone company representatives, the tariffs are available for public inspection at your telephone company office.

3) Phone patches should not be made solely to avoid telephone toll charges. Patches should never be made when normal telephone service could be just as easily used. The primary purpose of a phone patch is to provide vital, convenient access to authorities during emergencies. Operators should exercise care, judgment and restraint in placing routine calls.

4) Third parties (nonhams) should not be put on the air until the responsible control operator has explained to them the nature of Amateur Radio. Control of the station must never be relinquished to an unlicensed person. Permitting a person you don't know well to conduct a patch in a language you don't understand amounts to relinquishing control because you don't know whether what they are discussing is permitted by FCC rules. Make sure you have a copy of the latest list of countries with which the US shares third-party

traffic agreements. It's illegal to talk via Amateur Radio to nonhams in most foreign countries!

5) Phone patches must be terminated immediately in the event of any illegality or impropriety.

6) Station identification must be strictly observed.

7) Phone patches should be kept as brief as possible, as a courtesy to other amateurs; the amateur bands are intended primarily for communication among radio amateurs, not to permit hams to communicate with nonhams who can only be reached by telephone.

8) If you have any doubt as to the legality or advisability of a patch, don't make it. Compliance with these guidelines will help ensure that amateur phone patch privileges will continue to be available in the future, which helps the Amateur Radio service contribute to the public interest.

equipment. Dummy antennas are used to troubleshoot or to perform preliminary tune-up of your transmitter.

External Speaker

The small, low-power speaker in your receiver or transceiver may not produce high-quality audio output, especially at high settings of the volume control. Cabinet resonances may cause a vibration at certain audio frequencies and this can be distracting. Most internal speakers direct the sound upward or downward, rather than toward the operator. This isn't an ideal situation and it makes weak-signal reception difficult. An external speaker can solve these problems.

Choose an outboard speaker that contains a sound-reproduction unit with good midrange frequency response. This type of speaker rejects low frequencies and higher audio frequencies. The midrange response helps to minimize high- and low-pitched QRM, while allowing desired audio frequencies to be heard without attenuation. A hi-fi style speaker is suitable if you don't mind hearing interference from stations nearby in frequency.

If your receiver produces, say, ½ W of audio-output power, select a speaker that can accommodate 5-10 W of audio power. This practice ensures minimum audio distortion at high volume levels. Too small a speaker will sound tinny and distorted when it's operated near or beyond its power rating. Buy a speaker whose impedance matches your receiver's audio output line. This is usually 4 or 8 Ω. Check your operating manual for this specification.

Your external speaker is most effective if you locate it in the same plane as your head when you're seated. Place the front of the speaker so that it points toward your face. If you center it on a line to your nose, both ears will receive equivalent sound from the speaker.

You can build your own outboard speaker cabinet. Use thick wood and glue the seams of the box to prevent cabinet vibrations. I use ½- or ¾-inch particle board or plywood for my enclosures. I sandpaper the enclosure, then coat it with polyurethane varnish to provide a smooth finish that will accept wood-grain contact paper. I use window screening (painted black) for my speaker grilles. You can make your own grill cloth (if you wish to add this over the screen) from a piece of burlap or a section of a face towel (your choice of color). If you decide to use an external speaker, pick one that's 6-8 inches in diameter; large speakers produce better audio quality.

Antenna Switch

If you plan to have two or more coaxial-cable fed antennas, you'll want a coaxial antenna switch. You can build or buy a suitable unit. This switch permits you to change antennas quickly, and you won't need to unscrew and reattach cable connectors each time you use a different antenna.

Antenna switches are sometimes located remotely on a tower to permit using a single feed line for an HF-band beam antenna, a 6-meter beam antenna and perhaps a 2-meter vertical. Remote switches are operated from your ham shack using a dc control voltage. Manually operated antenna switches are used when the switching function takes place inside your radio room. Mine is mounted on the wall, just under the sill of my ham shack window. It can be easily reached from my operating position. My switchbox is homemade. I use an Ohmite power tap switch that I bought from a surplus dealer for $1. It's a single-pole, six-position switch. I housed it in a small aluminum Minibox.

24-Hour Clock

Local time is seldom used in Amateur Radio. Logbook and QSL card entries are referenced to the 24-hour system of Coordinated Universal Time (UTC). UTC is sometimes referred to as Z (Zulu) time, a shortcut term based on UTC being measured from 0° Longitude (the longitude of Greenwich, England, which accounts for the older term, GMT, for Greenwich Mean Time). A digital or analog clock in your shack is useful, especially if you plan to work DX and do a lot of QSLing.

Linear Power Amplifier

I've saved this "biggie" accessory until last. You may never own and operate a big RF power amplifier. I know many hams who have achieved DXCC, Worked All Zones (WAZ) and Worked All States (WAS) with modest power—even QRP. They don't feel that high power (QRO) is necessary for any form of MF, HF or VHF amateur operation. They keep the household power bill low and minimize the chances of causing RFI and TVI. I was licensed for 20 years before I used more than 100 W of power on any amateur band. Despite the lack of

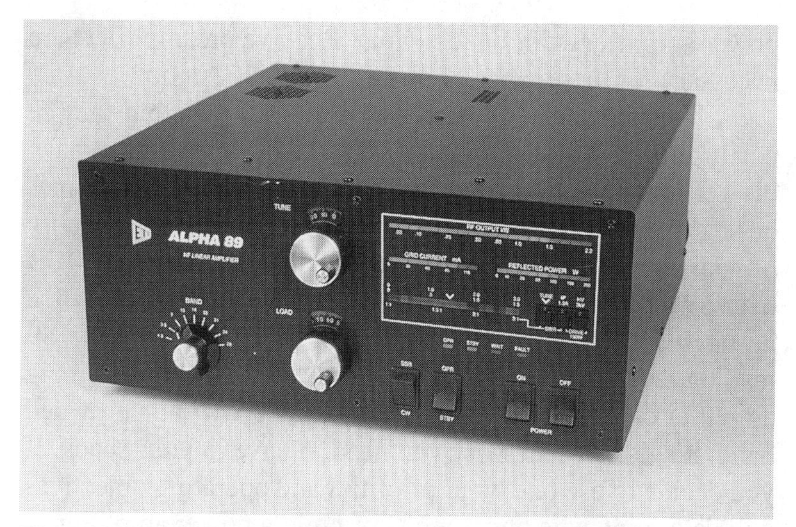

Fig 9—Linear amplifiers allow high-power operation—up to 1500 W—on the amateur bands. You don't need an amplifier to enjoy most Amateur Radio activities, however: On-the-air experience will bring you more success and satisfaction than simply cranking up the power.

high power, I talked to amateurs the world over without difficulty. High-performance antennas were more efficient, and less costly and troublesome than a linear amplifier!

Unfortunately, some amateurs erroneously think that full legal amateur power is essential, even for talking across town. The rationale is, "I'll override the QRM," and "I'll keep the channel open with my big signal." Philosophically speaking, amplifiers have always been used (or were supposed to be) when low power wouldn't provide solid communications. An example of this principle is when you operate on 75 meters when solar flares cause signals to be weak. An amplifier with up to 14 dB power increase can often assure solid copy when a 100-W signal may be lost in the noise. If you're interested in VHF or UHF DXing via tropospheric propagation, meteor scatter or earth-moon-earth (EME, or moonbounce), a linear power amplifier is all but essential. Receive preamplifiers are necessary for these types of operating activities, too.

Linear power amplifiers are useful at times, but they're costly and space consuming on your desk. Most of them have blower motors for tube cooling and fan noise can sometimes be heard on your voice signal. This doesn't improve the quality of your signal. I've observed SSB signals that peaked at 20 dB over S9 on my S meter when the operator was talking. When he paused between words, the meter registered S9. The S9 reading resulted from amplifier-blower noise! Although a big amplifier can make you "the bully of the town" in a crowded band, it's not an accessory you need to have in your shack. If you decide to use one, your priorities and operating objectives will determine whether you buy or build a power amplifier.

Summary

I've discussed the more common accessories that may find a place in your ham shack. My advice is that you consider the need and utility of these extra gadgets before you buy them. The fewer things you have to adjust and use will make your first days as an amateur much less complicated.

Chapter 9

DXing and Contest Operating

You've entered an arena of brisk competition. Chasing DX (communicating with distant amateur stations) is a sport for the stalwart amateur. The same is true of the contester who seeks the coveted ranks of the winners.

To succeed in this highly competitive pastime, you have to learn basic procedures. Your skill and success as a DXer or contester will be founded on experience and careful observation of how the game is played. Spend plenty of time listening to how polite, experienced operators conduct themselves in DX "pileups" (many stations calling a single DX station at the same time). It's important to know when and how to insert your call sign when the DX operator stands by for his next QSO. Timing is important when you chase DX or operate in a contest. Courtesy should be foremost in your mind during these activities. Those who transmit on top of a DX station's transmission—and many unskilled hams do—deprive other operators of their acknowledgments and signal reports. Besides, a DX operator can't hear your signal if he's transmitting. Be courteous and call him after he stands by. Make sure he's finished with the previous QSO before you call him. Your turn will come if you're patient.

There will be times during your entry into the world of DXing and contesting when you'll feel discouraged. Don't give up! Your confidence and skill will increase as you develop good operating techniques. Don't be the "bully of the frequency" with illegal power and discourtesy. Amateurs who demonstrate the greatest skill, courtesy and patience are eventually recognized by their peers for operating properly and may be nominated as a select member of the exclusive ARRL A-l Operator Club, for which an impressive certificate is issued.

Most successful DXers and contesters equip their stations with above-average antennas. Although you can win contests and earn certificates with dipoles and verticals, your chances for success increase when you install gain antennas, such as Yagis or cubical quads, atop tall towers with large rotators. Full-sized verticals with quality ground-radial systems can aid your cause on 160, 75 and 40 meters. VHF and UHF contesting champions often use stacked arrays or beams.

Equipment Needs

I know hams who have garnered the ARRL DXCC award (for confirming contacts with 100 countries) with low power and simple antennas. It took them two or three years and considerable skill and patience to accomplish this extraordinary feat. I also know amateurs who worked 100 countries in two or three days, using gain antennas high above the ground. This is not unlike the classic story about the tortoise and the hare; the slow mover eventually reached his goal. A case in point is my colleague, Wes Hayward, W7ZOI, who earned his DXCC award with homemade equipment that produced less than 25 W of output power. His antennas were made from wire (dipoles) and weren't high above the ground. It took Wes a few years to earn his award, but patience made it possible.

VHF/UHF operators work toward various levels of the League's VHF/UHF Century Club (VUCC) awards. These are presented to amateurs who submit proof of having contacted stations in the required number of grid squares (see the sidebar "What's Your Grid Square?") on each of several bands above 50 MHz. Although many hams have done it, if you think it's a snap to work stations in 100 or more grids on 50, 144, 222 or 430 MHz, you're in for a surprise. It's a challenge and an exciting pursuit.

On HF, you might start with 100 W of output power. This will make the job easier than if you try to use QRP equipment. Modern HF transceivers are in the 100-150 W power class, so let's assume that you're equipped in this manner. VHF/UHF transceivers usually put out much less power, such as 10-50 W. If you add a linear amplifier, you can get up to about 10 dB of additional signal strength. This is slightly less than two S units of gain over your "barefoot" signal. Most amateurs never use or need a powerful linear amplifier because the extra 10 dB may not be noticeable under most conditions, but one or two additional S units can sometimes spell the difference between being copied solidly or being lost in the noise and interference. FCC rules and good operating practice call for an amateur station to use high power only when it's needed.

There are many antennas you can buy or build for use at your DX or contest station. It would be impractical to describe them here. Get a copy of the *ARRL Antenna Book*, which describes many suitable DX antennas in detail. If you prefer a

> **What's Your Grid Square?**
>
> One of the first things you'll notice when you tune the low end of any VHF band is that most QSOs include an exchange of grid squares. Grid squares are a shorthand means of describing your general location anywhere on Earth. (For example, instead of trying to tell distant stations that, "I'm in Canton, New York," I tell them, "I'm in grid square FN24kp." It sounds strange, but FN24kp is a lot easier to locate on a map than a small town.) Grid squares are coded with a 2-letter/2-number/2-letter code (such as FN24kp). This handy designator uniquely identifies the grid square and your exact location in latitude and longitude; no two have the same identifying code. There are several ways to find out your grid square identifier. The ARRL offers a grid-square map of North America, a World Grid Locator Atlas and a program for PC-compatible computers (*GRIDLOC*).—*Mike Owen, W9IP*

plain-text approach to building homemade antennas, get *W1FB's Antenna Notebook* from your local retailer or directly from ARRL HQ.

Your Receiving Setup

There's an age-old saying in Amateur Radio: "If you can't hear 'em, you can't work 'em." Keep this in mind if you want to be a successful DXer. Not only must you have an antenna that can gather weak signals, you need a sensitive, stable receiver. Modern transceivers and receivers have wide dynamic range (the ability to handle strong signals without overloading) and low noise figures. They have stable local oscillators and offer selectivity options that enable you to separate signals that are close together. A wide dynamic range

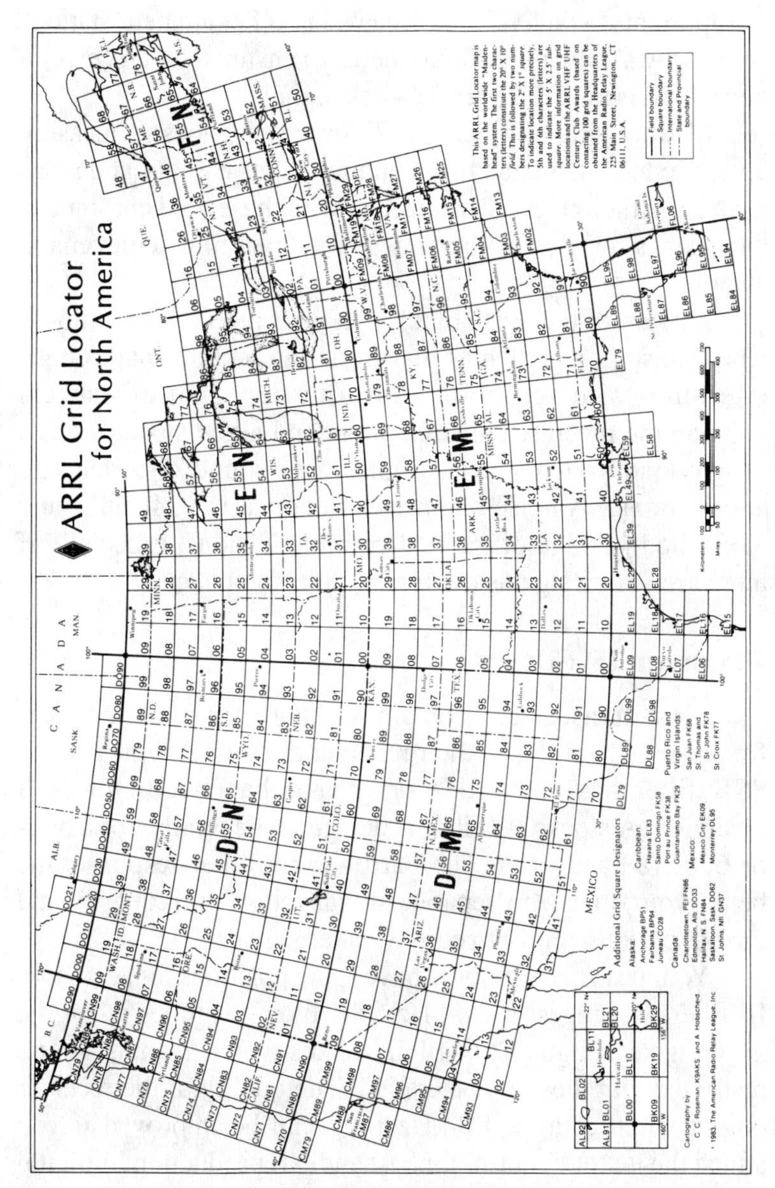

Fig 1—The ARRL Grid Locator Map. You can use a map like this to determine your present grid location and the grid locations of other stations. You can order your own copy of this 18×12-inch map for the wall of your shack for $1 from ARRL HQ.

is important if you live in an area where other amateur stations are nearby. A low noise figure helps to ensure that a receiver's internal noise won't override weak signals.

Most older equipment (1975 or earlier) doesn't measure up to modern standards for dynamic range, stability and noise figures. Although you can succeed with these "old chestnuts," they can slow down your quest for operating-achievement awards.

Beware of the noise blanker in your receiver. Use it only when pulse noise, such as that from automobile spark plugs, interferes with reception. Unfortunately, noise blankers degrade the receiver dynamic range and allow strong signals to overload your receiver. Signals can sound distorted and appear broad. A noise blanker can make a DX pileup sound like a hodgepodge of garbled communications, especially if there are strong signals on or near your frequency.

Receiver IF Filters

Most transceivers contain stock (basic) intermediate-frequency (IF) filters in their receiver circuits. A typical stock SSB filter has a 2.4-kHz or greater bandwidth for SSB reception. A stock CW filter is designed, generally, for a 500- or 600-Hz bandwidth. The narrower the filter response, the better your chance to separate signals close in frequency and the better your ability to copy weaker signals.

Your transceiver should have spare positions for accessory IF filters. I suggest a 1.8-kHz SSB filter for times when the going gets rough. This filter makes voice signals sound restricted, but they can be copied. The quality of the receiver output when using a 1.8-kHz filter can be improved if you adjust the IF SHIFT control for best audio reproduction. A digital signal processing (DSP) filter is a wonderful device to use for tailoring your reception for optimal response characteristics.

I prefer a 250-Hz CW filter and a selectable 600-Hz filter. I use the wider filter for most CW work, but when I copy a weak signal in a noisy band or when interference is a problem, I use the 250-Hz filter. The narrower CW filter tends to "lift" weak signals above the noise threshold, which can mean the difference between solid copy or no copy at all. I don't consider the addition of the two narrow filters a frivolous expenditure for a DX or contest operator.

Using an Audio Filter

Audio filters were covered extensively in an earlier chapter, but they're worth mentioning again. This accessory can be useful if you don't have narrow SSB and CW filters in your receiver. In terms of overall receiver performance, an audio filter falls short compared to a good IF filter, but it can greatly enhance reception of weak signals. I recall an experience I had during a DXpedition on Montserrat, a West Indies island in the Caribbean, which would have been a flop without the audio filter I took to that DX site. Atmospheric noise that close to the equator is extremely bad. The QRN registered 10 dB over S9 night after night on 160 and 80 meters. My receiver had only a 600-Hz IF filter for CW. Every time I called CQ on 160 meters, I could hear CW stations answer me, but I couldn't pull their call signs out of the noise. I placed my audio filter in the line and set it for a narrow bandwidth. Previously unreadable signals became Q5, even though the noise could still be heard plainly. I had the same experience while operating CW on 80 meters. My VP2MFW operation would have been a failure were it not for the audio filter.

If you purchase or build an audio filter, make sure it has a variable bandwidth control and a variable-peak-frequency feature. Fixed-tuned audio filters are satisfactory for routine operating and are simple to use, but you'll fare better with adjustable filters during difficult periods of reception.

Split-Frequency Operation

We mentioned this topic earlier, but we didn't focus on split-frequency operation for DX operating. Split-frequency operation permits you to transmit on one frequency while listening to another frequency. The DX station may be operating in a part of the band where you aren't allowed to transmit, but listening for answers in your part of the band or the DX operator may be in your portion of the band, but listening one or more kHz above or below his operating frequency for those who answer his call. You must use split-frequency operation in these situations.

There are many ways to operate "split." Some amateurs use a separate receiver with a transceiver. They listen to the DX station with the outboard receiver and transmit with the transceiver. This arrangement permits you to monitor your operating frequency without changing the transmitter frequency. Keeping tabs on what's happening on your own frequency is helpful with respect to timing when you're trying to break through a DX pileup.

You may use an outboard VFO with your transceiver to permit split operation. Many modern transceivers have two built-in VFOs, and this is ideal for split operation. If the frequency difference between transmit and receive is less than 2-3 kHz, you can use your RIT or XIT controls to cover the small frequency split.

You'll seldom succeed when calling a DX station on his frequency. Many calling stations tend to do that and the resulting bedlam of interference on the DX station's frequency makes it difficult for him to sift a single call sign from the tangle. You'll fare better by calling him 1-2 kHz above or below his frequency. The interference will be of lower magnitude away from his frequency and he'll probably find you by using his RIT. I've never listened on my operating frequency when

I was on a DXpedition; I always listen at the fringes of the pileup—usually 1-2 kHz above my frequency.

Snagging the DX Station

Most CW DX stations operate in the Extra Class parts of the US amateur bands. You'll usually find them at 3500-3515 kHz, 7000-7025 kHz and 14,000-14,025 kHz. This should provide an incentive for you to upgrade your license class later on, but don't despair. Many DX stations are heard in the General-class CW bands. You can work them in the bands that don't have Extra Class segments, such as 10, 12, 17 and 30 meters, but the most intense DX operating seems to occur at the low end of each CW and phone band. This phenomenon has prevailed for decades. I don't know why.

Suppose you wish to communicate with a European station. How should you proceed? First, choose a part of the day when the band is usually open to that part of the world. Next, spend time tuning and listening. This will give you a grasp on band conditions and the level of DX activity at that time of day. Once you establish that the band is suitable for DX QSOs, you might call CQ or answer the CQ of a DX station. A general CQ will probably net you a stateside reply because US amateurs will assume that you're simply looking for someone with whom to chat. You'll fare better by sending a directional CQ, if DX is your sole objective. In this situation, you can CQ as follows:

CQ CQ CQ EU DE NØETY K

The EU signifies that you prefer to have a QSO with someone in Europe. Although it's acceptable to transmit CQ DX CQ DX CQ DX DE NØETY K, some hams object to this procedure and feel that a directional CQ is more appropriate because a general CQ DX may yield a QSO from any corner of the world.

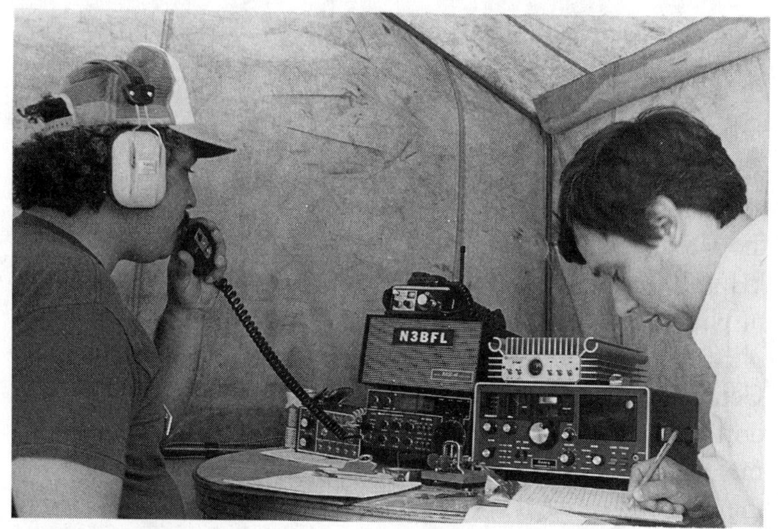

Jim Kelley, N3BFL, operates 2-meter SSB with Gene Marcus, W3PM, logging.

Calling CQ DX makes it clear that you don't want to talk to hams within the borders of your country.

Maybe you want to talk to someone in a particular country. Try using a directional CQ. Suppose you want to talk to someone in Spain. Your CQ can be structured as follows: CQ EA CQ EA CQ EA DE N1HAT $\overline{\text{KN}}$. The EA is for the Spanish prefix and the $\overline{\text{KN}}$ (sent as one character, DADIDADADIT) indicates that you prefer to have only an EA station respond. Amateurs use $\overline{\text{KN}}$ during ragchews with friends. This tells other hams that you don't want to get into a multistation QSO while you're chatting with a friend.

Answering the CQ of a DX Station

Things can be complicated if, as a newcomer, you answer the CQ of a DX station. The situation worsens if the operator is in a rare country (such as a small country with few hams).

Rare countries are in such demand that a "pileup" of stations is sure to begin when the operator from that land calls CQ. Not only will you hear US stations calling him, but hams from all over the world will become part of the pileup. A timid operator may be afraid to get involved in the resulting bedlam!

QRZ means "Who's calling me?" A DX station operator may be unable to copy a single call sign the first time he stands by because of intense interference or he may only get part of a call sign. If so, he may come back with QRZ AW? with AW being the suffix of, for example, W1AW. At this point, all other callers should stand by for the station or stations with the AW suffix and allow the contact to take place.

The DX operator may not only be unable to extract a call sign from the pileup, but may not even get a suffix or a part of one. In this case, he'll wait until the roar from all the calling stations subsides, then call CQ or QRZ again. Eventually he'll recognize one caller and initiate a QSO. The loudest, cleanest signals are the ones he'll acknowledge first. Be patient because eventually you'll be heard as the layers of louder signals diminish, assuming you're using average power and a modest antenna. It may take 15 or 20 minutes to be heard by the DX station, but don't give up.

This is a good example of when to call the DX station above or below his frequency, as discussed in the previous section. Use your RIT or XIT to place your signal at the edge of the pileup.

Avoid long replies to the CQ of a DX station. Not only is this type of conduct apt to brand you as a lid, but it's not effective for obtaining a reply from the DX operator. It's better to be terse when calling and if you aren't recognized the first time around, wait and try again.

Sometimes a DX station, hoping to make the pileup more manageable, will request calls only from stations in certain areas or just those with specific prefixes. If this happens, play by his rules. If you give in to the urge to call out of turn, it can disrupt the whole arrangement and invite others to do the same. Then the pileup quickly reverts to a blur of noise and interference, a few self-proclaimed "band police" may turn up to chastise those who have transgressed, until soon the fun and satisfaction trying to work the DX station is gone… as may be the DX station itself. As always, the advice is to be patient, courteous and a bit cagey. A skilled operator will eventually work almost any DX he really wants.

You may want to adopt the "tail-ending" technique when calling a DX station. It consists of simply giving your call sign once, such as DE KB1AFX, just after the DX station completes a QSO with another station, but before it calls CQ or QRZ again. As a skillful tailender, you'll make sure that the DX operator has finished his transmission before dropping in your call sign. You need not include the call sign of the DX station. This saves time and minimizes the pileup magnitude. The only time you must give the other station's call sign is at the end of an exchange of third-party traffic with a foreign country. Careful, considerate tail-ending is an acceptable practice.

Avoid being "windy" when you contact a DX station that has others waiting to work it. Stick to the basics of the QSO, such as your name, QTH, his signal report and perhaps QSL information. Sign off quickly and allow the next person to work him. If the DX operator wants to ragchew with you and if no one else is waiting to work him, you can go ahead and discuss other topics. Some DX station operators never ragchew, while others prefer to chat for a while. You'll be able to recognize which type of person the DX operator is by paying attention to his operating style and the nature of his conversations. Follow the DX ham's lead.

Tail-ending takes skill and timing to be successful. I generally wait until the pileup subsides somewhat before I drop in my call sign. This ensures that I have less competition and the chances are that my signal will stand out better than if I were to give my call sign at the start of a pileup. This procedure isn't as necessary if you reply to the DX station a couple of kHz off his frequency, where the number of signals is smaller. I've broken large pileups with 2- and 5-W CW rigs. Good timing and calling off frequency made this possible.

It's seldom a matter of who has the most power and largest antenna when it comes to getting the attention of the DX station. No amount of hardware can make up for a lack of operating proficiency, and the most skilled operators can make a contact under seemingly impossible conditions. Always consider skill the most important component in your amateur operations. Excellent operating techniques separate the world's greatest ham operators from the lids and the amateurs who are just "boys with expensive toys."

Don't let momentary frustration spoil your enjoyment. If a pileup is simply too huge or band conditions are plain bad, give it a rest. Take a break and try to work the DX station later or on another band...or another day. No QSL card, award or certificate is worth the cost of your blood pressure or sanity!

The DX Ragchew

If you have the good fortune of finding a DX station operator willing to chew the rag, cherish it! Don't be disappointed if the person is unable to talk for a long period. He may have limited knowledge of your language and may not understand all that you say to him. In this situation, it's vital that you speak slowly and clearly. Slang expressions and regional sayings (such as, "I bent over backwards for her.") are seldom understood by those who live in foreign countries. You'll discourage a foreign ham if you don't stick to clear, simple English. Avoid speaking rapidly on SSB. A foreign ham with a limited grasp of English will have difficulty understanding a fast talker. It's similar to trying to copy CW at speeds beyond your capability. Don't send CW at a speed faster than that being sent by the DX operator you're in a QSO with.

This is a good rule no matter who you're communicating with; it's a courtesy all hams should extend.

If ragchewing with DX operators is something you enjoy, try to plan your contacts accordingly. It's more likely that you'll find smaller pileups and less competition for a DX operator's attention if you come across him on a weekday morning, after most US hams have gone to work or school, calling CQ or answering your CQ DX call. There are DX operators who prefer ragchewing over running strings of quick contacts because they cringe at the prospect of being deluged under a pileup—and the subsequent blizzard of requests for QSL cards—whenever they fire up their radios. If you can tell that the DX ham wants to have a nice chat, settle down and enjoy the pleasure of getting to know about him, his country and his interests. This is your chance to spread international goodwill, and to experience a foreign culture and perspective from the comfort of your shack. Many warm, enduring friendships have formed between hams in different countries who have never met except via the amateur bands.

Avoid Certain Topics

There are extreme differences in religion, politics and culture around the world. It's easy to offend others if you force your philosophies on them. If you broach certain delicate topics, you may inspire the DX contact to sign off hurriedly. Suggestive remarks and vulgarity have no place in Amateur Radio and this type of talk may be especially offensive to a ham from another land.

There are many acceptable things to discuss during a DX QSO. Almost any hobby has universal appeal. You can begin a satisfying conversation by talking about your line of work, your families and Amateur Radio in general.

Be truthful at all times. Don't inflate the signal report you give to the DX operator. He's more interested in knowing how well his station is doing abroad than in being flattered. Inflated signal reports won't hasten or ensure the receipt of his QSL card.

Contesting

Contesting tends to "divide the men from the boys." You'll need internal fortitude if you decide to become a dedicated contest operator. Competition is fierce, there are long, tiring hours of operating and your disposition can get ragged after an all-night stint at your key or mike. If it's done right, contesting can build your operating skill and confidence. Try this exciting pastime.

The rules of courtesy are as applicable to contesting as they are to DX operating. Patience and timing are vital to your success as a contester. The QRM can be unbearable in a busy band during a contest and your resolute nature may be at stake if you allow the racket to get the best of you. But take heart—you can do it.

There are two kinds of contest stations: (1) single operator and (2) multioperator. If you're the only participating operator at your station, you're in the single-operator class. This means that you must conduct all of the related activities. You can't operate the equipment while someone else does the logging, nor may you operate the contest radio while watching a local *PacketCluster* station for spotting. A multioperator setup has several operators. A multioperator station may employ two or more operators who share the work, using one transmitter, with several antennas and perhaps additional receivers for "spotting" multiplier stations. This is called a multisingle operation. Many multiop stations use separate transceivers and antennas to cover several bands simultaneously. This makes it a multioperator, multitransmitter (multi-multi) station.

Only one call sign is used for a multi station. The operators can be assigned to shifts. This allows time for eating, resting and sleeping, and for taking turns logging while another mans the controls.

Most contests provide for score multipliers, depending on the type of contest. During ARRL Field Day (only), you can rack up extra points for using low power, "alternative" energy sources (batteries, solar cells, waterwheels or windmill generators, etc) and for copying the W1AW message. In some contests, you earn additional points for working foreign stations or ARRL Sections. Read the rules for details before the contest. Accurate logging and strict adherence to the time limits of the contest are imperative to avoid being disqualified.

How to Operate

There's no better guide for contesters and DXers than the ARRL *Operating Manual*. Get a copy of this book because it provides an in-depth view of all amateur operations. Meanwhile, here's an overview of how to operate a contest.

Your most productive operating method is to select a clear frequency for your contest operation. Stay put on that frequency and let the other stations come to you. The highest scores are usually earned when this procedure is adopted. Remain on your frequency and send CQ CONTEST and your call sign. This is called "running stations."

A less productive technique is to keep moving around different frequencies and replying to others who are calling CQ CONTEST. This is called the "hunt-and-pounce" technique. Your strategy is to call every station you hear and to be especially alert for stations that count as multipliers or for extra points. It's less likely that you'll run up the highest score if you operate in this manner. You'll lose time tuning from one frequency to another. The station you call may be a frequency hopper: You may call her and obtain no reply because she's already moved to a new frequency.

The frequency-hopping scheme is okay if you're a casual contest operator and aren't attempting to roll up the winning overall score. That itself is an honorable pursuit. The sport of contesting needs new participants to provide contacts—and points—for the regular, "serious" contest stations. You may enter a contest only to provide points for those who are competing with their peers. Frequency hopping is fine for this purpose, too. It's fun to participate in contests without being a "serious" competitor. You can take advantage of the opportunity to collect new states for your WAS award or new countries for your DXCC award. Casual contest operating is useful for determining how competitive your signal is. If you're often answered on your first call, and if your message is copied the first time, you can feel confident that your equipment and antennas are working well.

The odd thing is that many amateurs who give contesting "a quick try, just this once" find themselves hooked. In a few cases, a modest effort may even pay off in a certificate in a category that doesn't attract many entrants. For example, if you live in a state with few hams, you could find yourself winning a certificate in a less-competitive category. For example, in 1992, a ham who wasn't an experienced contester surprised himself—and his ARRL HQ staff colleagues—by winning first-place in the first ("1") call area in the single-operator, single-band (20 meters), low-power, unassisted category in the *CQ* Worldwide WPX SSB contest. It took him about six hours of operating on a Sunday afternoon, using a common solid-state transceiver, an untuned horizontal loop of wire strung 25-35 feet above his house and trees, and an inexpensive antenna tuner. Later that year, he invited a friend to join him in his first stab at the ARRL November Phone Sweepstakes, and their unassuming multioperator entry racked up enough contacts to earn them commemorative participation pins and "Clean Sweep" coffee mugs (and a few good-natured, jealous grumbles from his fellow staff members).

Certificate Hunting

There are countless awards you can earn for operating achievements. The ARRL sponsors several. You can earn a Worked All States (WAS) certificate upon proof of your contacts with hams in the 50 US states. This requires sending your confirming QSL cards to ARRL HQ for inspection and verification, or you may have your cards checked by an HF Awards Manager. The cards are returned to you when the certificate is issued. The same procedure is used when you apply for the DX Century Club award (DXCC, for working 100 countries). There's also the Worked All Continents (WAC) award and the VHF/UHF Century Club (VUCC), mentioned earlier. The Worked All world Zones (WAZ) is issued by *CQ* magazine.

You may wish to pursue the County Hunter's certificate. To earn this award, you must provide confirmation of contacts with hams in various counties in the US. Some amateurs devote years to this challenging pastime, with the goal of making and confirming a contact with an amateur station in each of the 3076 counties in the US.

Another award that may interest you is the A-1 Operator Club. It's sponsored by the ARRL and a certificate is issued after two anonymous A-1 Operator Club members nominate you for club membership. Membership is restricted to amateurs who exhibit the most skillful, proficient and courteous on-air techniques. You can't apply for membership and if you're fortunate enough to be chosen, you'll never learn who your sponsors are. It's a deep secret and an extreme honor. A-1 Ops are the elite among their peers!

Code proficiency certificates are available from the ARRL if you successfully copy messages sent by W1AW at a variety of code speeds during scheduled Qualifying Runs. Complete information about awards and contests is provided in the ARRL *Operating Manual.*

Don't rely on the signal reports you receive during a contest. Most operators will say that your signal is 59 (SSB) or 599 (CW), although your signal may be a mere 549. The 599 reporting system is used for expediency in a contest. It allows the memories in keyers or computerized logging/keying/voice announcing systems to be programmed for a specific signal report (and other information) and this saves time. (In this regard, I'm somewhat a maverick. I always report the other station's signal strength and readability as it really is.)

Contest Logging

You need an accurate clock and log sheets for contest operation. For most contests, a "dupe sheet" is essential for each band of operation. This extra contest form has the amateur prefixes for each US call area and many DX countries printed on it in numerical order. There are blank spaces under the prefixes where you list the suffixes of the stations you work. If you're in doubt about whether you've already worked a calling station (or one you wish to call), a quick check of the dupe sheet reveals the data you need to prevent working the same station twice on a given band. Many operators use computers to perform the function of dupe sheets. As we learned earlier, duplicate QSOs on a given amateur band can cause you to be disqualified in a contest. Some contests allow working a station once on phone and once on CW on the same band, however.

If you use a computer or keyboard keyer for contesting, you may be able to program it to serialize your contacts and record the time of the QSO. This is an easy task for your computer with one of the many contest software packages available. Hard copy can be taken from the computer to provide the data you need when filling out your contest forms or you can print out a summary sheet and submit it with your entry on a computer diskette, via a telephone modem or even on the Internet.

Contest forms for ARRL-sponsored events are available free from ARRL HQ. Information about forthcoming contests is published in *QST* a month or two in advance. ARRL publishes the *National Contest Journal* (*NCJ*) six times per year. Other organizations that sponsor contests include *CQ* magazine, QRP clubs, radio clubs and many overseas radio societies.

VHF/UHF Contesting

Amateur Radio contests test your ability to work the most stations in different geographical areas on the most bands during the contest period. Contests give you a chance to evaluate your equipment and antennas and to compare your results with others. In most VHF/UHF contests, each contact is worth a certain number of points. You multiply your point total by the total number of different grid squares (multipliers) to calculate your final score. The only restrictions in these contests are that contacts through repeaters (and satellites) don't count, and the national 2-meter FM calling frequency, 146.52 MHz, is off limits for ARRL contest QSOs. SSB and CW are the most popular contest modes, but you can have a lot of fun with FM, too. During the first hour or two of a VHF contest, contacts may come fast and furious. Then the pace slows, as operators prowl the bands looking for new stations.

Who Can Enter?

Most VHF/UHF contests are open to any licensed amateur who wants to participate. The ARRL sponsors the major VHF/UHF contests (see Table 1), and specific rules, descriptions of the categories and entry forms are available free from ARRL Headquarters. You don't have to be an ARRL Member to participate in these contests, nor are you required to submit your logs, although doing so helps the ARRL Contest

Branch verify QSOs that others claim. VHF/UHF contests feature a variety of categories among which you can choose. For single operators (those operating without assistance), entry classes in ARRL contests include all-band, single-band, QRP portable and one for rovers (those who operate from more than one grid square during the contest). The ARRL *Operating Manual* is a good source of information on selecting an entry category and what it takes to go QRP portable or roving to different grid squares during a contest

When and Why?

ARRL VHF contests are held throughout the year, with emphasis on the warmer months to encourage hilltop operation, during favorable weather. Outside of that, the ARRL VHF/UHF contest program is designed to take the best advantage of propagation enhancements that usually occur at certain times of the year. For instance, the June VHF QSO Party almost always occurs during periods of excellent sporadic-E propagation, giving you an opportunity to enjoy long-distance contacts on 6 and 2 meters. The first documented sporadic-E contact on the 222-MHz band was made during a June VHF QSO Party. As shown in Table 1, the major VHF/UHF contests are the January Sweepstakes (SS), June and September VHF QSO Parties, the *CQ* WPX-VHF contest, August UHF Contest and the VHF/UHF Spring Sprints. Except for the Sprints, these events encompass many bands each. The January Sweeps and June and September QSO Parties are the most popular of them all, and each permits activity on SSB, CW and FM on all amateur frequencies from 50 MHz and up. The UHF contest is slightly different than the other contests described so far. The major difference is that only contacts on the 222-MHz and higher bands are allowed for contest credit. The Spring Sprints are single-band, four-hour contests held over a several-week-long period. These

Table 1
Major VHF Contests

Contest	*Bands*	*Time*
ARRL VHF Sweepstakes	50 MHz and up	Varies according to Super Bowl weekend
Spring Sprints	One per band	April/May
June VHF QSO Party	50 MHz and up	2nd full weekend
CQ WPX-VHF	50 MHz and up	July
August UHF Contest	222 MHz and up	1st full weekend
September VHF QSO Party	50 MHz and up	2nd full weekend

Check the contest rules ahead of time and make sure you know the exchange requirements. Some contests, for example, require that you give each person a contact serial number, beginning with 001. Your first contact would be 001, your second 002, your fifteenth 015 and so on.

short contests provide opportunities to test new locations or equipment.

VHF/UHF Contest Exchanges

As in all Amateur Radio contests, there's a standard exchange of information between stations. In many cases, the exchange consists of your grid square location and a signal report. (Signal reports are optional in ARRL-sponsored VHF/UHF contests.) You can hunt for stations to call or find a clear frequency and call CQ yourself. Here's an example:

CQ CONTEST, CQ CONTEST FROM N1BKE, NOVEMBER 1 BRAVO KILO ECHO, N1BKE

[N1BKE calls CQ to initiate contest contacts.]

N1BKE FROM ALFA ALFA 2 ZULU

[AA2Z responds, using phonetics so there's no confusion about his call sign.]

AA2Z, YOU'RE 59 IN FN32 FROM N1BKE

[N1BKE responds, giving a signal report and his grid square.]

N1BKE FROM AA2Z. ROGER. YOU'RE 57 IN FN20, OVER

[AA2Z confirms receipt of N1BKE's information and sends his own.]

ROGER, THANKS FOR THE CONTACT

[N1BKE confirms AA2Z's report and continues…]

CQ CONTEST, CQ CONTEST FROM N1BKE, NOVEMBER 1 BRAVO KILO ECHO, N1BKE, QRZ?

[With the exchange made and confirmed, N1BKE resumes calling CQ for the next contact.]

When to Be Where

You'll find lots of 6- and 2-meter activity during VHF contests. On SSB, most stations stay near the calling frequencies of 50.125, 50.20, 144.20, 222.10 and 432.10. On CW, look between 80 and 100 kHz above 50, 144, 222 and 432 MHz. (Six meters offers less CW activity than the other VHF/UHF bands.) There's plenty of fun contesting with FM; read the accompanying article from *QST*'s FM column.

FM Contesting

(Adapted from the FM column in January 1994* QST*)
By Brian Battles, WS1O

Where were you during the last VHF contest? Did you miss the ARRL Spring Sprints in April and May? The ARRL UHF Contest in August? The September or June VHF QSO Parties? Will you be ready for the January VHF Sweepstakes?

The standard answers are, "I'm not a contester," "I don't have an all-mode rig for those bands" or "I don't have time to stay on the air all those hours." Fortunately, none of these standard answers are valid reasons to ignore the events. You may say, "The only ham radio activities I like are handling traffic and chatting with friends on the local repeater." It's funny how that's not much different from operating in a VHF contest.

"I'm Not a Contester"

Who is? Only a person who elects to participate in a contest. And even then, you aren't really known as a contester by anyone else unless you submit an entry log. If your main interest is ragchewing, helping with public service activities, organizing emergency communications or handling traffic on the local repeater, you can also have fun by trying your hand at a VHF contest. Making a contact is easy. You simply exchange your call sign and grid square. Unlike the big HF contests, an FM contest QSO can be much more leisurely. There's rarely any QRM to fight, and because there are fewer stations in range, there isn't the frantic pace HF contesters must maintain. By the way, some of the best contesters come from the ranks of traffic handlers, which makes sense because contest operation requires the ability to copy the other station accurately and efficiently.

"I Don't Have an SSB or CW Rig"

You don't need one. Every ARRL VHF/UHF contest lets participants use FM. You simply can't use repeaters. There aren't any special multipliers for working DX stations or having a 1000-W amplifier feeding a stacked array of Yagis. Each FM simplex contact with a neighbor is worth just as many points as a CW contact with a station 1000 miles away. That hand-held 2-meter FM transceiver can net you enough points to make a strong showing in your Section—if you go to the trouble of using it. How much trouble? Pick a simplex frequency (see below) and listen for—or call—"CQ contest."

"I Don't Have Time"

If you plan to be on the air over a contest weekend, you have time to join the fun. Simply exchange the necessary information and write it in your log. There's no minimum number of operating hours or contacts you have to make, no bonus points for staying awake all night, no special awards or certificates for climbing Pike's Peak or Mt McKinley, operating from a submarine or auto gyro, or standing on your head. (Although some of those could be great opportunities to take a photograph and send it in to ARRL HQ for Up Front in *QST*!) You can take a stab at the contest while sitting comfortably in your shack, living room or car for an hour or two. Grab a snack and a mug of coffee or hot chocolate, a couple of pencils and a log sheet. If you don't have a blank contest log, use a plain piece of paper. You can always copy the info onto a standard contest log afterward, if you decide to "officially" enter. If you prefer, boot up a contest-logging program on your computer and let the silicon do the thinking. You can get such software by mail order, by downloading it from local telephone BBSs or national online services, or ask for a copy from almost any contester you know.[1]

"I Never Hear Anyone Using FM During Contests!"

If no one else is calling "CQ Contest," then you may as well do it. You'd be surprised at how many others might pop out of the woodwork to make a QSO. An almost foolproof strategy is to recruit your friends and members of your club to join you to give each other contacts. This can lead to hundreds of "easy" points and give your contest log a shot in the arm. The January VHF Sweepstakes features three categories of competition for ARRL-affiliated clubs. Pool your logs as a club entry and share the glory. There's even a certificate for the high scorer in each club. Other hams you convince to try it may also become regular contesters, too, and perhaps you can form the nucleus of your own team. (As an added incentive, if you can get 25 people you know to get on the air during the ARRL September VHF Sweepstakes, and each of you works everyone else, you'll all be eligible for participation pins!) If you want to make it easier to convince friends or club members to participate, ask them to call or listen on a particular frequency at the top of each hour from, say, 1 PM to 6 PM. That way, they'll be more likely to find someone and make a contact or two, rather than randomly turning on the rig, hearing nothing and giving up. There are well-known frequencies to go hunting on: Try the following frequencies: 144.9-145.0, 146.49, 146.55, 146.58, 147.42, 147.45, 147.48, 147.51, 147.54, 147.57; 223.5; and 446.0 MHz. Don't use the 2-meter national simplex frequency, 146.52 MHz, for calling or soliciting contacts.

"What's in it for Me?"

Contests put a lot of hams on the air. The FM simplex frequencies that may be normally quiet most of the time will usually be fairly busy. You'll get a better idea of how effectively your station functions, what kind of range your station is capable of spanning, how propagation and seasonal conditions affect your station, and what neighboring stations you can hear.[2] You might get your call sign or your club's name into the contest results write-up in *QST*. In some Sections with a smaller active ham population, your modest effort could earn a certificate or plaque! You can make new friends on the air, who probably don't live too far away. You might discover a propagation "opening" and experience the thrill of working someone hundreds of miles away with your hand-held transceiver. You'll hand out QSOs for points to other contest operators. (As a newcomer, you'll quickly learn how welcome you are, because your call sign will be very noticeable to regular local VHF contesters who may be frustrated by always hearing nothing but the same bunch of stations during every contest.) Most important, it's almost certain that you'll have fun.

[1]Excellent programs are readily available and inexpensive. If you're just starting out in contesting, you may be better off with *freeware,* which costs nothing to use, or *shareware*, which you may use free for a trial period, but requires a nominal registration fee if you plan to use it regularly. If you have a modem, but don't know where to start looking for contesting software, try the ARRL HQ BBS (203-666-0578) or commercial online services like CompuServe, NVN, America Online and GEnie. If you have access to the Internet, there are sites that carry ham radio files for downloading. Commercial contesting software is available from commercial vendors that advertise in *QST* and the *National Contest Journal* (published by the ARRL).

[2]I've been astonished by the possibilities of long-distance communications on 2-meter FM. For example, there was a band opening one evening not long ago when I made simplex contacts from my home in Connecticut with stations in Cape Hatteras and in Maine, and I was even called by a station in Panama City, Florida, although I couldn't complete a two-way QSO with that station. I wasn't using an exotic rig, antenna, power amplifier or preamp: my trusty ICOM IC-2GAT hand-held transceiver (about 8 W output) was feeding a 5/8-wavelength whip antenna on top of the chimney of a two-story house.

Summary

You may never try some of the activities discussed in this chapter. You may prefer to participate in casual activities, such as ragchewing, net operation or traffic handling. You might be more interested in experimenting and technical achievements than in chasing DX or building a competitive contest station. You might enjoy building and testing equipment more than operating it. Networking may be your personal interest, and you could be happy spending all of your Amateur Radio time at a keyboard. Perhaps ATV suits you, and you'll be happiest pointing a camera or performing in front of one. There's something for everyone when it comes to station activities. In any event, I encourage you to try the activities discussed in this chapter—you may discover that DXing and contesting are what you've been looking for in ham radio.

Chapter 10

Logs, QSL Cards and Record-keeping

Although maintaining a station log might seem like a waste of time, it isn't. In the past, the FCC required amateurs to keep a timely and accurate station log of all transmissions. The data required in the logbook was the time of activity, transmitter power, mode of operation, the date and the call sign of the station worked. You even had to log periods when you operated mobile and portable—at one time, you even had to log every time you just called CQ. Logging is no longer mandated and some hams don't use logbooks. Nonetheless, there are advantages to keeping a station log.

It's easy to do so because today most hams have a home computer, and there are dozens of logging software products available for almost any type of PC. They cover the gamut from no-cost freeware to low-cost shareware to complex, do-it-all retail packages. Most keep track of your contacts and can produce reports with contacts sorted by bands, countries, states, dates and more. Sophisticated programs alert you of a country you need for an award as soon as you type in a prefix—or when a spot comes over the local DX *PacketCluster* you're connected to—translate a prefix to the name of the country, and even provide instant "pop-up" utilities to improve your QSO with the distant ham, such as previous contact information, beam headings, short- and long-path distances, his local time, data on his country's culture, graphics with maps and flags, and perhaps a screenful of key phrases in the DX operator's native language!

If that's a bit much or if you don't have a computer, a good, old-fashioned notepad or preprinted logbook will suffice. A log has a place for general comments about your QSO and standard data columns. If you maintain a day-to-day log, you'll have a valuable record that contains the names of the operators you chat with, their call signs, when you talked to them and details about their equipment. The general-information column on your logs provides information you'll need the next time you talk to those people. Furthermore, your logbook will contain your signal report and that of the other station. This data can be helpful if you or the other operator decide to try a new antenna. Cumulative signal reports for a given time of the day are useful for propagation studies if you become interested in that facet of the hobby.

As an example, suppose you called CQ and received an answer from W1CKK. She greeted you by saying, "Hi, Sam, it's been two months since we had a QSO." At this point your mind can't summon her name or location. Her call sign sounds familiar, but you can't recall the details of your last conversation. You may feel embarrassed because of your poor memory. If you had kept a log of your activities, it would be simple to skim quickly through it and extract her name and

QTH. Without the logged data, you must ask her to give you her name and location. Think about how nice it would be if during your first transmission you could say, "Hello, Jean. I haven't heard you on 75 meters since our last QSO last summer. Are you still using your inverted V and homemade 100-W transmitter?" She'll be pleased that you've "recalled" the pertinent information about her station and can call her by name.

There are other good reasons for logkeeping. Suppose you've used a 40-meter loop antenna for several months. You've logged numerous signal reports from hams you speak to frequently. Therefore, you have a good idea about how your average signal strength has been over a period during which your loop was used. Yesterday however, you decided to erect a 2-element 40-meter Yagi and you're eager to learn if there's been an improvement in signal strength. Now's the time to contact stations you've worked previously, obtain a signal report, then compare it with the average or singular report you got from that station in past weeks or months. You can enter this new data in your logbook for similar future use. It requires many tests with a number of previously worked stations before you can develop a new average signal strength. This method isn't particularly accurate because band conditions vary from hour to hour, day to day and week to week. The correct way to make meaningful comparisons between antennas is to have them both erected and well isolated from one another. Rapidly switching from antenna 1 to antenna 2 will enable another station to observe changes in loudness and S-meter readings as you switch between antennas. It's important, therefore, to recognize that antenna tests over a period of days or weeks (without the antenna-1/antenna-2 quick change) provide only an average reference. Later, after you've been a ham for eons, it's fun to be able to recall nostalgically which transceiver and antenna you were using at a given period in your hamming

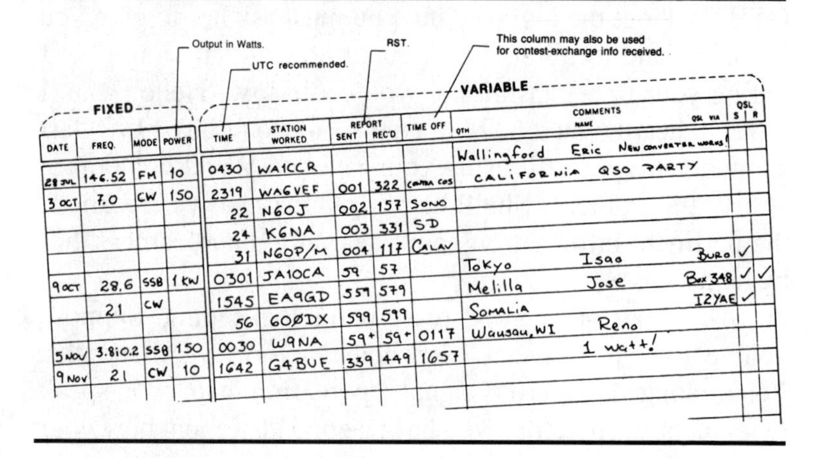

DATE	FREQ.	MODE	POWER	TIME	STATION WORKED	REPORT SENT	REPORT REC'D	TIME OFF	QTH	NAME	COMMENTS	QSL VIA	QSL S	QSL R
28 JUL	146.52	FM	10	0430	WA1CCR				Wallingford	Eric	New converter works!			
3 OCT	7.0	CW	150	2319	WA6VEF	001	322	Contra Cos	California QSO Party					
				22	N6OJ	002	157	Sono						
				24	K6NA	003	331	SD						
				31	N6OP/M	004	117	Calav						
9 OCT	28.6	SSB	1 kW	0301	JA1OCA	59	57		Tokyo	Isao		Buro	✓	
	21	CW		1545	EA9GD	559	579		Melilla	Jose		Box 348	✓	✓
				56	6OØDX	599	599		Somalia			I2YAE	✓	
5 NOV	3.810.2	SSB	150	0030	W9NA	59+	59+	0117	Wausau, WI	Reno				
9 NOV	21	CW	10	1642	G4BUE	339	449	1657			1 watt!			

Fig 1—A logbook can be useful for many reasons, but most hams keep a logbook simply to recall the stations they've worked over the years.

days. Writing down details can preserve a rich lode of your personal memories.

Logbooks are important if you become a DXer. It normally takes months or years to obtain confirmation of 100 DXCC countries. An accurate record of your DX QSOs will prove useful if you need the QSL card for a country you lack. For example, you may have worked three or four 8P6 stations on Barbados, but the first one or two didn't send a QSL card upon receiving yours. You can now search your log sheets and request a QSL card from one or more of the remaining 8P6 stations. Include with your card a self-addressed return envelope and an international reply coupon (IRC) that covers the cost of postage. Alternatively, you may route your QSL card through the DX country's QSL bureau. Don't pass over this concept too quickly; there are many DX fans who started as 2-meter FM ragchewers, but later in life found the urge to upgrade, pick up some HF gear and stalk the bands for exotic foreign prefixes. The ones who never kept a log when they

Fig 2—A 24-hour clock like this one (digital clocks are also available) can help you think in terms of 24-hour time—the time hams use around the world.

started out may curse the day a tropospheric band opening gave them a QSO with a startled ham in Cuba or Clipperton on 2 meters!

The 24-Hour Time System

I keep my log, make schedules and fill out QSL cards in universal coordinated time (UTC). I suggest you do the same. UTC uses a 24-hour format. It's recognized worldwide. UTC helps when making schedules and getting QSL cards from hams abroad. You translate your local time and other hams do the same. In the US, we add hours to local time to get UTC as follows: AST/EDT 4 hours, EST/CDT 5 hours, MST/PDT 7 hours and PST 8 hours. Add 10 hours in Hawaii to get UTC. Local time is a poor choice because most other hams, especially DX operators, use UTC exclusively.

It takes some practice to feel at home with UTC or local 24-hour time. After you've adapted to it, you'll have to convert your local time (EDST, CST, etc) to UTC. You'll need to keep track of seasonal time changes to make accurate conversions. For example, when you change from Standard Time to Daylight Saving Time, you have to take this into account. A 24-hour station clock makes timekeeping a simple task. Get one if you aren't adept at using UTC. These clocks are available in analog or digital formats. A quartz movement assures accurate time over a long period, but weekly or monthly calibration against WWV, WWVH or CHU is wise if you want to ensure high accuracy for your clock. I use WWV at 10.000 MHz for my calibrations. The Canadian time-standard station, CHU, is at 7.335 MHz.

Logging DX

Logbooks are useful for non-DX logging, too. Suppose you're seeking your ARRL Worked All States (WAS) award. Although you've kept track of the states you've worked on a sheet of paper or by filling in the outlines of states worked on a map, you still need an accurate record of each station's call sign and the date, time and band of operation when you fill out your QSL card. The same is true when progressing toward the Worked All Continents (WAC), Worked All Zones (WAZ) and other achievement awards. You can buy logbooks from ARRL HQ for a nominal fee.

It may take you several years to accumulate enough confirming QSL cards for your DXCC application. Suddenly you recognize that you lack only one or two cards to provide the 100 confirmations you need for the DXCC award. If you've kept an accurate log over the period of DX chasing, chances are that the last one or two countries you need have been worked. Scanning the log sheets will often reveal QSOs for which you didn't send your QSL card. Most operators indicate on the log-sheet whether they sent a card to the station in question. Notations are made to indicate that a confirming QSL card was received from the DX station. This is a vital part of logkeeping if you're an award seeker. So although we're no longer required to keep a station log, the advantages of having one are numerous.

Another Logbook Advantage

Even if you're only interested in VHF/UHF, FM repeaters or packet as you start out in Amateur Radio, there are technical and sentimental reasons to maintain a log; and sometimes it may serve as "insurance." There can be a time (horror of horrors!) when you're cited by the FCC for improper operating procedures or for having a defective signal. In this instance, you'll receive a show-cause notice (known as a "pink ticket"), to which you must respond in a specified time frame. The notice will contain data concerning the band of operation, date, time of day, call sign of the station being worked, plus a rundown concerning the nature of the infraction. You may refer to your station log (1) to ensure that you were on the air at the specified time and (2) if on that day you had changed equipment or antennas. This data may provide a clue as to why your station received a citation.

Upon checking your logbook, you may learn that you weren't on the air the day the event was supposed to have occurred. Someone may have miscopied another call sign as yours, or worse, you may be the victim of a bootlegger. On rare occasions, an unlicensed person invades the amateur bands and for the lack of a call sign, may use an assigned call sign. Yours may be the one a scofflaw elects to use on a particular day. His misuse of the equipment or improper on-the-air conduct may cause you to receive the dreaded pink ticket. I can speak with conviction on this subject, because I've received three pink tickets in 40 years, for offenses committed by bootleggers using my call sign. In each incident, I sent a photocopy of my logsheet for the day of the alleged offense to the FCC engineer who cited me. I included a polite letter explaining that I was at work or asleep when the infraction took place. The charge was dismissed when the FCC engineer received my information.

I've been cited for TVI and RFI that I didn't cause. The interference in my neighborhood took place while I was at work and on another occasion it happened while I was on vacation 200 miles from home. My logbook was invaluable for proving my innocence.

I recall a nasty case of illegal operation when I lived in Detroit during the early 1950s. A local amateur wasn't liked by one of his peers (jealousy?). The victim was a prominent DXer who had a loud signal worldwide and who had the most commanding signal in Detroit on 20 meters. The disgruntled amateur tape recorded the innocent ham while he was calling CQ during voice operation and replayed it several times outside the US portion of the phone subband. The innocent victim received three pink tickets from the FCC! Fortunately, he kept an accurate log. When he submitted copies of the logsheet for that day to each FCC office, he was exonerated. A subsequent investigation revealed the real culprit.

QSL Cards

QSL cards are personal—they reflect your personality and tastes, and they're your tools for obtaining the QSL cards of other hams. It's important that your QSL card not contain artwork or printed matter that could offend others. I've seen cards that contained profanity and suggestive artwork. Others had political slogans that could be offensive to those who don't embrace the same philosophy. If your QSL card is inappropriate for some reason, your chances of receiving a confirming QSL card are minimized. QSL cards should reflect the top-quality things about Amateur Radio.

Your QSL card should have ample space for entering data such as the date, time, frequency, mode of operation and the other station's call sign. Additional space may be used for listing the type of equipment and antenna you're using, although other operators may not be interested in your equipment lineup.

Attractive artwork helps to make your QSL card appealing. Most amateurs who display QSL cards select the most attractive ones for their walls. You may wish to design your own card. If you send the rough or finished artwork to a QSL card printer, he'll usually prepare your cards to order for a special fee. Inspect the free QSL card samples printers offer (see the *QST* Ham Ads for suppliers). You may find a design that appeals to you. All you need do after making a selection is provide the printer with a layout of the printed matter you wish to have on the card.

I've always avoided buying my QSL cards from printers who sell inexpensive cards for those with limited budgets. These cards aren't attractive. Many of them are in use and your low-budget card will look just like dozens of others being circulated. These cards are okay in an emergency, but you'll be happier if your cards are distinctive and attractive.

I mentioned earlier that your QSL card reflects your personality and that the nicer the card, the better your chances for garnering that elusive DX QSL card you're seeking. You may want to have a photograph of yourself (seated at your rig) printed on your QSL card. This adds a nice personal touch

Fig 3—You'll need valid QSL cards to apply for many awards and certificates, but if you're like most hams, you'll also want to start a collection of cards as a memento of the stations you've contacted.

and pleases those with whom you chat frequently. Knowing what the other person looks like is always an interesting experience—our mental images of how other hams look are usually wrong!

If you're a photographer and have your own darkroom, you may want to manufacture your own photographic QSL cards. I once did this and the cards were excellent.

Tips for Obtaining Confirmation

It's important to send a self-addressed envelope with your QSL card. This reduces the cost for the person who will send his card to you. If you're seeking a QSL card from someone in your country, place a postage stamp on the return envelope. International Reply Coupons (IRCs) may be sent with your card to hams in other countries. Be sure to include enough IRCs to cover his airmail postage cost.

Some amateurs, in the interest of saving a few pennies, send their QSL cards at postcard rate. I prefer to send and receive QSL cards in an envelope. This prevents damage to the card and eliminates the possibility of having the post office canceling stamp obliterate important data on the card, such as the date of the QSO or your call sign. Spend a few more pennies and send your cards in envelopes. The other person will appreciate your tender loving care.

Direct QSLing to stations abroad will shorten the turnaround time for QSLing. If you route your cards through the QSL bureaus, it may take months to receive the card you need.

Resist the temptation to send money with your QSL card. Money isn't necessary if you use IRCs. Some hams have been known to include cash as a persuasive prod toward getting a rare DX QSL card. I've received cash "gifts" from amateurs who wanted my DXpedition QSL cards. I always return their money when I confirm the QSO. It's wise to keep in mind that cash offerings may insult a person in another country. He may regard the gesture as a form of bribery. Sometimes you may feel inclined to voluntarily include a token fee to help offset the expense of a particularly difficult, rare DXpedition, but that's your choice. There's no justification for having to pay for a QSL card.

If you wish to add spice to the QSL pot, send the DX station a photograph of your shack or a picture postcard of a significant area of your town or state. I've sent postcards many times and was pleasantly rewarded by receiving a similar card from the DX station, which depicted a point of interest from his region. After a particularly friendly or memorable chat, it can be a nice touch to send your QSL card and include a unique memento, such as a sticker, business card or local tourism pamphlet.

The Outgoing QSL Service

The ARRL operates an Outgoing QSL Service for Members. You may take advantage of this by saving your outgoing QSL cards until you have about a pound of them to send abroad. You sort the cards alphabetically and numerically before shipping them, in accordance with country prefixes. In other words, all DL cards are in one pile, ZL cards in another pile, etc. After you've sorted the cards, put them in one stack in alphabetical order. Don't separate the cards for individual countries with rubber bands, paper clips, etc. The ARRL bureau workers place incoming cards in pigeonhole slots until there's a sufficient number for each foreign call area to justify a bulk shipment. The League charges a modest fee for its QSL service ($2 per pound in 1994). This saves you considerable money over mailing your cards separately via First Class or airmail. There are special arrangements for official League-affiliated clubs, too. Send a self-addressed, stamped envelope (SASE) to ARRL HQ for complete information about the Outgoing QSL Service.

QSL Managers

Some DX operators delegate the chore of personally sending their QSL cards to those who request them. They use volunteer amateurs who serve as QSL managers. For example, suppose you work 8P6CC on Barbados. As he signed off with you, he said to QSL via W1UED. This means that your card should be sent to W1UED, rather than to 8P6CC. This arrangement reduces your postal costs and it hastens the receipt of the card from 8PSCC. The DX operator periodically sends a copy of his logsheets to his QSL manager so that the manager can verify the contacts of those who want cards. As I mentioned earlier, include an SASE with your card when you send it direct to a QSL manager.

Sending QSL Cards to ARRL HQ

You'll be sending bundles of QSL cards to ARRL HQ for awards sometime during your amateur experience. These are cherished cards that often can't be replaced. Shipping your cards by any carrier is a gamble because mail and packages are lost from time to time while *en route*. Personal delivery of your cards is the safest way to transport them, however inconvenient it may be. What alternatives are there? I ship my QSL cards by registered mail. I've had problems with delivery when using the US Parcel Post system. I avoid using that carrier for items that can't be replaced. Pay a little extra for special handling and allied services that help to ensure safe and fast delivery. Bundles of QSL cards have been lost in the mail and it can be a dreadful experience to learn that months or years of QSL card collecting have been in vain. UPS and Federal Express are my second choices. They'll provide the peace of mind that comes with knowing that your cards arrived safely. Include with your cards sufficient funds for return service following ARRL verification of your QSOs. Each card is checked carefully to ensure its validity.

Other Records to Keep

Record the serial numbers from all of your station equipment. Put a photograph of each piece of gear in your file. Make notes on the distinguishing characteristics, such as a scratch on a cabinet, a missing or mismatched knob, chipped finish and so forth. This will help the insurance carrier to identify items that are stolen if your ham shack is burglarized or destroyed by fire, flood or other catastrophe. Photographs of your station equipment are helpful to law-enforcement agencies when identifying stolen property that's been recovered. Include the serial numbers and pictures of your mobile gear, computer and other valued items, too.

Develop a list of items you lend to amateur friends. Record the item loaned, the recipient's name and the date he borrowed the item. This may seem like a strange recommendation, especially when it concerns friends, but it's easy for the borrower to forget that he borrowed an item and you may forget who you loaned something to. This includes books, station accessories, tools and other items you don't want to lose. I've lost power supplies, books, tools and even a linear power amplifier because I failed to keep track of who borrowed the material. Now I keep an accurate log that shows who borrowed what and when.

Summary

The guidelines offered in this chapter should make your ham radio experience easier and more enjoyable. There are topics I didn't cover, but most common situations are covered here or in the ARRL *Operating Manual*.

Chapter 11

Obtaining Accurate Information

Incorrect technical or operating information can cause inconvenience and grief. It's easy to pick up misleading or false information on the air. A well-meaning amateur may think he has the answer you need, but he may have gathered his "facts" from an unreliable source. ("Everybody knows that. . .") There are technical myths that have haunted hams for years. An untruth may appear in an Amateur Radio magazine article because the author thought he was conveying accurate data. Some hams avoid reading about an unfamiliar subject—it's easier to ask a friend on the air if he knows the answer to a question. We live in a busy society that dictates finding expedient solutions to little problems.

Erroneous information is easy to generate. Suppose I decide to experiment with an antenna I've never used before. I spend two weeks running tests in the field and on the air, and conclude that its performance is excellent. I performed my field tests with makeshift test gear and my on-the-air tests weren't conducted scientifically: I didn't erect a reference dipole against which to compare the new antenna. My lab notebook has many notations that I consider to be factual. Now I go on the air and laud the performance of this great antenna. Other hams ask questions and take notes so they can try the antenna. They tell others about the antenna and might even quote some of my numbers incorrectly. This isn't uncommon when stories are repeated. I'm so impressed with the antenna that I write a magazine article about it. Now the false data is in print forever! Generations of amateurs may read this article and believe everything I said, however false it may be. People tend to believe anything that's published, so my questionable comments are cast in concrete as a factual reference. Had I done my homework and been thorough during my antenna testing, this unintentional hoax wouldn't have been perpetrated.

Everyone hears technical chatter on the air, at club meetings and at hamfests. How can you determine what's true and what's false? Separating truth from fiction isn't easy. You can't rely on the information source just because the ham who provides advice has been licensed for years. On the other hand, a newly licensed amateur may be a fountainhead of accurate technical data. It depends on the person's technical education, background and attention to detail. I know technical whizzes who have no formal education in electronics. These self-taught experts never make a technical statement without first researching the subject in a textbook, and they conduct practical investigations to verify what they've learned.

If you have a friend who's an electronics engineer or technician, ask him for advice when you need it. Another reliable source of this information is your local ARRL

Technical Coordinator (TC). You can get the address and phone number of your TC by contacting your ARRL Section Manager (SM), whose name, call sign, address and telephone number is on page 8 of *QST*. If you write to your TC for technical assistance, include a self-addressed, stamped envelope (SASE) for the reply. The postage expense for these volunteer field appointees can be high in a busy year. The cost of envelopes adds to the expense, too.

If you phone your TC for advice, do it during the day or early evening. Daytime calls on weekends are good, too. Most ARRL field appointees don't like to receive phone calls at mealtime or late in the evening when we're relaxing or watching a favorite TV program. Try to be brief on the phone: The TC may have as many as five or six calls in an evening, and would enjoy relaxing after a day at work. Some TCs list their work phone numbers for your convenience. If you can't reach your TC by phone, ask your question in a letter. Perhaps he or she also has an electronic mail address that's even more convenient.

Reference Books

You'll find answers to nearly all of your technical questions in ARRL books. The *ARRL Handbook* and the *ARRL Antenna Book* have been the standard references for amateurs. The League has many titles that belong in your amateur library, such as the ARRL *Operating Manual* and *W1FB's Antenna Notebook*. The *ARRL Electronics Data Book* and the *FCC Rule Book* should be a welcome additions to your collection. There's an extensive list of books you can get from ARRL HQ. These books are advertised in *QST*, so keep an eye out for listings.

Owning a technical library is not, in itself, the solution to your need for technical information. You need to establish a pattern of regular study if you are to find answers to your questions. Schedule regular time for "book larnin'." This regimen will prove valuable if you decide to upgrade your amateur license. You'll understand the exam questions and answer them correctly through study, rather than guessing or memorization. Guessing at answers causes many examinees to fail the tests. Use the ARRL's *FCC Rule Book* and the appropriate ARRL *License Manual* as you prepare for your next license exam. If questions in the *License Manual* perplex you, find the topic in the *ARRL Handbook* or another League book. Study that section until you understand the question.

Don't overlook the value of maintaining a file of *QST* back issues. The ARRL's monthly membership journal keeps you abreast of rule changes and technical developments that may have a bearing on the nature of the questions asked by the FCC at exam time. Furthermore, there may be articles that don't interest you when you receive the current issue of *QST*, but you may develop a keen interest in those articles months or years later. As you develop new skills and gain technical knowledge, old subjects can become ones of current interest. The construction projects and theory articles in *QST* are carefully checked by the ARRL HQ staff before the articles are published. This is your assurance that what you read in *QST* and ARRL books is accurate. Keep an eye on *QST* Feedback items because an occasional drafting error can occur or an author may discover that he provided a wrong part number or left a term out of an equation after an article is published. Corrections are printed under Feedback. Clip these corrections to the original article. This saves time and minor headaches later. Check your town library if you don't have a particular back issue of *QST*. Many libraries maintain a file of *QST*s.

ARRL Technical Information Services

The ARRL Technical Information Service (TIS) is a valuable membership benefit that can help you find answers to technical questions. The first place to look for a technical answer is in ARRL publications. If you call or write to the ARRL TIS staff, they will usually refer you to a book or to a past *QST* article. It is usually easier and faster, and much more educational, to look up the answer yourself. Each issue of *QST* contains a catalog that describes the League's current publications. Each book contains a comprehensive index. For magazine articles, check the December or January issue for the annual index.

The ARRL Field Organization is part of TIS. Each ARRL Section has a Technical Coordinator (TC) and/or one or more Technical Specialists (TS). These volunteers often offer personal advice about antennas, interference problems or other subjects that are best answered in person at the local level. Your ARRL Section Manager, listed on page 8 of any recent *QST*, can refer you to your TC or local TS.

Local clubs are an important technical resource. Most clubs have one or more friendly, experienced hams who can offer advice about antennas, rig selection or nearly any other topic. You might even find that one of the "old timers" has a *QST* or book collection that goes back many years, giving you ready access to a wealth of information. If you would like a list of clubs in your area, send an SASE to ARRL HQ with a request for a "Local Club State Printout."

If you have a question about a specific technical article, the author of that article is usually the best person to contact. Write the author a friendly letter, asking your questions. (Always include an SASE.) Most authors will send you a prompt reply. If you offer feedback, try to be positive, even if you disagree with the author.

ARRL HQ can answer technical questions, too. The TIS staff is a valuable source of information and research for questions that can't be answered by the TC or an article author. TIS offers, on request, a variety of information packages and bibliographies that will answer many common technical questions. The TIS staff can do research to locate an article on

TIS: 'Tis the Source You're Looking For

Another service provided by TIS is its computerized information server. If you have access to a computer that can connect to the Internet (directly or indirectly), you can request information from ARRL HQ's growing archives by sending e-mail. To get started, send a message to the following address:

INFO@ARRL.ORG

Each line of the message should contain a single command as shown below. You may place as many commands in a message as you want. Each file you request will be sent to you in a separate e-mail message.

Valid info commands:

HELP	Sends a help file
INDEX	Sends an index of files available from Info
SEND <FILENAME>	Sends filename, for example: SEND PROSPECT
QUIT	Terminates the transaction (use this if you have a signature or other text at the end of the message.)
REPLY <ADDRESS>	Sends the response to the specified address. Put this at the beginning of your message if your FROM: address isn't a valid Internet address.

Your message won't be read by a human, so don't include any requests or questions except by way of the above commands. Retrieve the ARRL-EMAIL-ADR file for a list of direct electronic-mail addresses of ARRL HQ staff members. Don't generate a personal message addressed to INFO@ARRL.ORG The system was set up to prevent failed mail from looping endlessly, to the frustration of system administrators. Mail sent to INFO-SERV@ARRL.ORG ends up in the "bit bucket."

Your FROM: field or REPLY-TO: field in your header should contain a valid Internet address, including full domain name. If your FROM: field doesn't contain a valid Internet address, the answer won't reach you. There's a reply function as a server syntax, as follows:

REPLY MAILADDR

Where MAILADDR is a valid Internet mail address (USER@DOMAIN, EG, JOHN_DOE@WHATNOT.COM). An invalid address generates an error. A wrong address results in nondelivery of your response. The address given in the REPLY command is the address to which all subsequent requests in the message will be sent. If an error message is generated, it will be sent to the last reply address given.

If you need help with the server, or if you have information files in ASCII text format that you'd like to archive on the ARRL HQ TIS server, additional information or updates for any of our files, or suggestions for improvements, contact ARRL Headquarters via e-mail or telephone 203-666-1541.

If you don't have a way to get mail into the Internet, you can use your computer and modem to dial the ARRL HQ telephone BBS at 203-665-0578. It allows you to search for and download hundreds of files, and to read and send electronic mail to ARRL HQ staff members and other BBS users.

a particular subject, find a company that sells "widgets" or explain a technical point in person by phone or letter.

TIS can supply photocopies of out-of-print articles or construction "templates" as mentioned in *QST* articles. Contact the hard-working Technical Department Secretary for information or help. Some back issues of *QST* are available from ARRL HQ for the current cover price. Contact ARRL Publication Sales to see if the issue you want is available.

Nontechnical information is available from the League on request. For instance, you can obtain operating aids, contest log sheets, W1AW schedules, award applications, maps and more.

Tips About Your Technical Inquiries

For your convenience, every *QST* article lists the author's name, call sign and mailing address. You may want to write to an author to have a statement clarified or to seek his advice about where to find parts for a project. If so, write a friendly, objective letter. If you don't agree with something he said, avoid introducing a negative tone in your letter. He may not answer your "sour grapes" letter.

Be specific when you ask questions. The author or TC may not be able to understand vague questions. For example, if you have a question about some part of his circuit, mention the figure number and page of his article (and the month and year of the publication). Give him the part number or numbers that relate to the circuit area of interest. If you're requesting a computer file or program offered in the article, include a formatted diskette and a self-addressed, stamped disk mailer.

Don't play "20 Questions." I frequently receive letters that contain 10 or more technical questions, and some have no relationship to an article I wrote. Authors receive numerous letters that relate to a specific article and there's seldom time to provide a two- or three-page reply. I've received letters from hams who tell me how much they enjoyed my article, only to change course and ask me to design a piece of gear for them. Authors and TCs don't want to get involved with projects like that. You'll do better if you stick to the theme of the article and be as brief as you can.

Suppose a *QST* author wrote an article about a high-power balun transformer and gave the details for building one. He probably received at least one letter that contained the statement: "Please explain how a balun transformer works." If I was that author, I'd probably tell the person that I lack the time to provide an in-depth treatment of the subject and suggest that the writer refer to the *ARRL Antenna Book*. Common sense is your best guide when you seek technical assistance.

ARRL Membership and *QST*

Your ARRL membership is a valuable resource. Some Members think of their membership dues as a *QST* subscription fee. This is not the situation. *QST* is a bonus for being a League Member. It's the official monthly membership journal of the ARRL. This can be equated to the publications of other membership societies, such as the National Rifle Association (NRA) or the National Geographic Society. *QST* keeps you abreast of current amateur developments worldwide. It provides information about FCC rule changes, spectrum allocation, contests, DX activities and a host of timely subjects. Without *QST*, you can get out of touch with events and techniques that pertain to your daily ham radio operations.

QST contains information about upcoming on-air events, hamfests and conventions. There are specialty columns devoted to VHF/UHF/microwaves, FM, club activities, ARRL Section news, official ARRL Board of Directors proceedings and satellite communications, and there's a section for beginners, the New Ham Companion. The commercial ads help you to stay up to date on new amateur gear and the Product

Reviews help you to make your choice when it's time to replace your old station equipment. A more subtle benefit of ARRL membership is the expenditure of part of your dues toward preserving Amateur Radio frequencies. The League is always in the forefront when it comes to defending the rights of amateurs. As an ARRL Member, you help to pay the cost of these important ARRL actions. You also have access to the ARRL Outgoing QSL Service, ARRL Equipment Insurance and other Members-only benefits. Every US ham should join and support the League.

Other ARRL Periodicals

You may be one of the many amateurs interested in packet radio, digital signal processing (DSP), microwaves, spread spectrum, satellites and advanced or theoretical experimentation. The ARRL produces a monthly publication, *QEX*, the ARRL experimenter's journal, which provides up-to-date, useful technical information for experimenters. The ARRL also publishes the *National Contest Journal* (*NCJ*) for contesters and DXers. *NCJ* has earned a reputation for providing top-notch contest reporting, useful techniques and equipment reviews. The *ARRL Letter* is a bimonthly newsletter that carries the latest news and information that relates all facets of League affairs and the world of Amateur Radio in general. It's sort of a mid-month supplement to *QST* to cover timely issues.

W1AW Bulletins

Late-breaking news of interest to radio amateurs is broadcast by ARRL Headquarters station W1AW, in Newington, Connecticut. Bulletins are transmitted on CW, SSB, RTTY (Baudot and ASCII), AMTOR and packet throughout the day on most amateur bands. The W1AW operating schedule and frequencies are published in *QST*. This is your primary source for up-to-date information about matters that affect Amateur Radio. W1AW broadcasts code-practice sessions on a regular basis. Various speeds are sent to provide copy practice for hams who want to upgrade their license classes.

Technical Seminars

The technical programs at hamfests and conventions are a fine source of reliable technical information. Most speakers are chosen because of their technical expertise and hands-on experience. There's usually a question-and-answer period at the end of the speaker's presentation. This is your opportunity to ask questions that pertain to the theme of the talk. Some speakers are willing to entertain questions that don't pertain to the talk they've given, so you may be able to squeeze in a few unrelated questions. Most speakers will decline to answer questions that concern subjects outside their field of expertise. They feel that it's better to provide no answer than to give advice that may not be accurate, so don't be disappointed if one of your questions isn't answered.

Summary

Use the recommended sources of reliable information spelled out in this chapter. If you doubt the accuracy of advice you receive over the air or in person, look up the subject in one of the ARRL technical books or consult your local TC.

You should never feel ashamed to ask a question. The day of the general practitioner in Amateur Radio has faded. This is an era of specialization in the electronics field and no individual amateur can be an expert in all facets of the hobby. Ask and learn! Some person or technical book can provide the answers to your questions, but never be afraid to ask. As the saying goes, there's no such thing as a dumb question; there are only people too dumb to ask!

Index

D

E

F

G

H

I

P

Q

R

S

W

X

Y

Z

About the American Radio Relay League

The seed for Amateur Radio was planted in the 1890s, when Guglielmo Marconi began his experiments in wireless telegraphy. Soon he was joined by dozens, then hundreds, of others who were enthusiastic about sending and receiving messages through the air—some with a commercial interest, but others solely out of a love for this new communications medium. The United States government began licensing Amateur Radio operators in 1912.

By 1914, there were thousands of Amateur Radio operators—hams—in the United States. Hiram Percy Maxim, a leading Hartford, Connecticut, inventor and industrialist saw the need for an organization to band together this fledgling group of radio experimenters. In May 1914 he founded the American Radio Relay League (ARRL) to meet that need.

Today ARRL, with more than 170,000 members, is the largest organization of radio amateurs in the United States. The League is a not-for-profit organization that:

- promotes interest in Amateur Radio communications and experimentation
- represents US radio amateurs in legislative matters, and
- maintains fraternalism and a high standard of conduct among Amateur Radio operators.

At League headquarters in the Hartford suburb of Newington, the staff helps serve the needs of members. ARRL is also International Secretariat for the International Amateur Radio Union, which is made up of similar societies in more than 100 countries around the world.

ARRL publishes the monthly journal *QST*, as well as newsletters and many publications covering all aspects of Amateur Radio. Its headquarters station, W1AW, transmits bulletins of interest to radio amateurs and Morse code practice sessions. The League also coordinates an extensive field organization, which includes volunteers who provide technical information for radio amateurs and public-service activities. ARRL also represents US amateurs with the Federal Communications Commission and other government agencies in the US and abroad.

Membership in ARRL means much more than receiving *QST* each month. In addition to the services already described, ARRL offers membership services on a personal level, such as the ARRL Volunteer Examiner Coordinator Program and a QSL bureau.

Full ARRL membership (available only to licensed radio amateurs) gives you a voice in how the affairs of the organization are governed. League policy is set by a Board of Directors (one from each of 15 Divisions). Each year, half of the ARRL Board of Directors stands for election by the full members they represent. The day-to-day operation of ARRL HQ is managed by an Executive Vice President and a Chief Financial Officer.

No matter what aspect of Amateur Radio attracts you, ARRL membership is relevant and important. There would be no Amateur Radio as we know it today were it not for the ARRL. We would be happy to welcome you as a member! (An Amateur Radio license is not required for Associate Membership.) For more information about ARRL and answers to any questions you may have about Amateur Radio, write or call:

ARRL Educational Activities Dept
225 Main Street
Newington CT 06111-1494
(203) 666-1541
Prospective new amateurs call:
800-32-NEW HAM (800-326-3942)

Notes

Notes

Notes

Notes

Notes

Notes

Notes

Notes

Notes

A
R
R
L

HELP FOR
NEW HAMS

PROOF OF
PURCHASE

FEEDBACK

Please use this form to give us your comments on this book and what you'd like to see in future editions.

Where did you purchase this book?

□ From ARRL directly □ From an ARRL dealer

Is there a dealer who carries ARRL publications within:

□ 5 miles □ 15 miles □ 30 miles of your location? □ Not sure.

License class:

□ Novice □ Technician □ Technician with HF privileges
□ General □ Advanced □ Extra

Name ______________________ ARRL member? □ Yes □ No

Call sign ______________________

Daytime Phone ( ) ______________________ Age ______________________

Address ______________________

City, State/Province, ZIP/Postal Code ______________________

If licensed, how long? ______________________

Other hobbies ______________________

____

Occupation ______________________

For ARRL use only	HELP
Edition	2 3 4 5 6 7 8 9 10 11 12
Printing	1 2 3 4 5 6 7 8 9 10 11 12

From ____________________

____

____

Please affix postage. Post Office will not deliver without postage.

EDITOR, HELP FOR NEW HAMS
AMERICAN RADIO RELAY LEAGUE
225 MAIN ST
NEWINGTON CT 06111-1494

.... please fold and tape ....